NORFOLK

IN THE

FOUR SEASONS

by

RON WILSON

Drawings by Robert Yaxley

The Larks Press

Typeset and published by
The Larks Press
Ordnance Farm House, Guist Bottom, Dereham
Norfolk. NR20 5PF
01328 829207

Printed by Lanceni Press, Fakenham.

May 1995

The right of Ron Wilson to be identified as the author of this book has been asserted by him in accordance with the Copyright, Designs and Patents Act 1988.

British Library Cataloguing-in-Publication Data
A catalogue record for this book is available from the British Library.

ISBN 0 948400 27 7

FOREWORD

Ron Wilson's book 'Norfolk in the Four Seasons' is the most concise, comprehensive and readable account of the flora and fauna through the annual cycle that I have had the pleasure of reading.

It is also a timely reminder, in these days of rapid and sometimes devastating rural change, that perhaps in Norfolk we have more to lose than any other county in England. Our landscape, environment and unique ecological diversity is also under greater threat than most.

'Norfolk in the Four Seasons' must be compulsive reading for everyone who cares about the future of the county.

Tom Cook
Sennowe Park

Vice-Chairman
Norfolk Wildlife Trust

CONTENTS

INTRODUCTION

From the spectacular sunsets over the Wash to the wildlife haunts of inland woods, Norfolk is a county of contrasts - and as the county of my birth, still holds a fascination which I have yet to find anywhere else in the British Isles.

For the wildlife enthusiast it is also a Mecca. Blakeney Point holds the record for the number of birds recorded anywhere in the British Isles; Broadland is home to one of our rarest butterflies, and Thetford Chase protects the now rare red squirrel.

The marshes of Stiffkey and Brancaster, although hardly welcoming to humans, provide a feeding place and home for an increasing number of birds - not to mention a wealth of plants and insects.

A county of contrasts, a county steeped in history which goes back to prehistoric times, a county which, in spite of modern developments, still has a charm all of its own - and a variety of habitats to match it.

It is because of the predominantly rural nature of Norfolk that there are so many different habitats which wildlife - both plants and animals - can exploit. But at the same time there are continuing threats to those plants and animals which make up the county's flora and fauna. And so, gradually, and almost imperceptibly in some areas, more and more wildlife haunts have disappeared either under the plough or beneath concrete.

It is because of this threat that organisations like the Norfolk Wildlife Trust - formed in 1926 and the oldest in the country - plays an ever important role in managing nature reserves as well as helping to ensure that wildlife continues to flourish in the county.

The work of the Trust is very influential, and it is a fact that when the first nature reserve was bought - before the Trust had been formed - the band of volunteers who contributed the cash, successfully outbid a sporting organisation! In the past when an appeal was launched for £100,000, this was exceeded by £30,000. It is heartening that even in these difficult times the vital work of the Trust goes on, thanks to the generosity of lovers of the Norfolk countryside. Whereas much of the

1

money comes from private individuals, there is an increasing input from commercial organisations, which helps to strengthen the ties between voluntary bodies and industry.

As an exiled Norfolkman I still have a soft spot for the county of my birth. Any profits from the sale of this book will be donated to the work of the Norfolk Wildlife Trust.

I am grateful to the Trust for their help and support, and to Paul Banham and Brian Webster for reading the manuscript and making valuable comments. However, any errors which remain are my own. Robert Yaxley has executed the fine drawings at quite short notice. Susan Yaxley of Larks Press has given me every help and encouragement and I am grateful for her willingness to publish this volume.

Ron Wilson
Daventry, Northamptonshire
April 1995

NORFOLK'S CHANGING COUNTRYSIDE

A farming revolution - Hedgerows - Verges and ponds -
Use of chemicals - Disappearing woodland and wetland -
Future outlook.

Not that many decades ago, the Norfolk countryside had hedges, ponds and woods almost everywhere, providing a wealth of shelter for the county's wildlife. Within a relatively short time, a revolution has taken place which has put our wildlife heritage in jeopardy. Hedges have been removed to increase field size, ponds have been filled in, and deciduous woodland continues to be uprooted. Where replanting takes place this is often with conifers. Even the most unobservant country-dweller cannot fail to have noticed the vast changes which have taken place in the Norfolk countryside. For the observant naturalist these changes are sweeping, causing alarm and anxiety.

Because of Norfolk's rural nature farming has inevitably affected the county's wildlife. Farmers were encouraged to produce as much food as possible. This led to intensive farming methods which meant that an increasing acreage of pastureland was ploughed up, and the almost indiscriminate use of fertilisers and herbicides affected the flora of any remaining pastures. Just over eighty per cent of the land in the British Isles is used for farming and forestry. Only about 0.2 per cent in England and Wales, and less than 1 per cent in Scotland has been set aside for nature reserves. It soon becomes obvious what effect farming has on any area.

Hedgerows form an important habitat for wildlife, and it has been estimated that the total area of hedgerows in Britain is almost twice that of the national nature reserves, and as a 'nature reserve' the importance of hedgerows cannot be over-emphasised. Although the floral composition of hedgerows is relatively sparse when compared to other habitats, the plants which occur are of great importance for the many animals which live there. For example, hawthorn supports over eighty different species of moths, and twenty-five of these feed exclusively on this plant. Estimates suggest that a million blackbirds use the hedgerow as home. Large scale destruction of this often underrated habitat has serious consequences for both flora and fauna.

Although some small farms still exist, they are generally a thing of the past, and larger units lead to bigger fields, and since more than two and half million acres of farmland have become 'urbanised' since the last war, hedgerows are 'wasted' land. Increased mechanisation also means that the hedgerows are barriers to efficient farming and are often removed.

Hedgerows date from various centuries, and some have been part of the country scene for at least a thousand years. Vast lengths of hedgerows were established between the sixteenth and nineteenth centuries as a result of the enclosures. In the first period of enclosures, fields were often divided by wooden hurdles, previously used for protection and for keeping in cattle. Many hedges probably arose accidentally along the line of the hurdles, where small shrubs started to grow, eventually developing into hedgerows. Where strip farming was practiced, rough vegetation grew along the lines of the dividing strips, again eventually forming hedgerows. Nowadays many hedges are flailed, often during the breeding season or in the autumn when the shrubs are laden with berries.

Most theories about the way in which wildlife uses the hedgerow are pure speculation. Little is known about the detailed flora and fauna of these areas, although much research has been carried out. Comparisons show that whereas one hundred acres of farmland may support anything up to two hundred and fifty pairs of birds of between twenty-five and fifty species, a similar area of moorland can support only ten to twenty pairs of three to ten species. Hedgerows therefore play a leading role in maintaining the diversity of bird life, since open fields support relatively few species - lapwing, skylark and partridge.

The smaller hedgerow plants which grow are influenced by the hedge's composition. For example, hedges composed mainly of ash have a poor flora, compared to a mixed hedge which is generally rich in plant species. The plants also influence the insect population, which in turn gives rise to an increase in larger species. At the top of the food chain are carnivorous animals, including foxes, owls and weasels. With the removal of a hedgerow a vast range of wildlife living in the hedge, and associated with it, is either destroyed or seriously affected.

In the past, roadside verges used to be cut once a year, in summer or early autumn. This activity increased, and at one point spraying with herbicides was also introduced. Fortunately roadside verge spraying has ceased, and the mowing regime has settled down to twice yearly, with other cutting mainly confined to danger areas like bends and crossroads.

Plants like hogweed and cow parsley, which provide plenty of shelter and food for insects and birds, are often cut down. But where they are left, their rampant growth often eradicates other smaller species, which in turn reduces plant diversity

Before the arrival of piped water, ponds were found in fields where livestock was kept. Because they are no longer needed ponds have been filled in, with work often being carried out in winter when there was little else to do on the farm. The destruction of farm ponds often resulted in many death; frogs hibernating in these areas were buried alive. Because frogs are cold-blooded animals they need to hibernate in water, burying themselves in mud during the autumn to avoid the cold winter weather. Although toads need ponds for breeding they are unlikely to be affected in the same way.

Spawning toads need relatively deep water, compared with the shallower areas which frogs select. Frogs appear to be more vulnerable than toads to changes in the countryside. Polluted pond water also affects breeding frogs, because they breathe through both their skin and their lungs. Recent incidence of infections affecting frogs is also giving cause for concern.

When chemicals are used on the land their effects can be catastrophic. Spraying has to be carried out in 'ideal' conditions, but this is seldom possible, and even on calm days drift from spray material can be considerable. Even with all the necessary precautions the effect on wildlife is often incalculable. In the past the most deadly chemicals were the organochlorines, which have now been replaced by less potent alternatives. These organochlorines were extremely toxic, remaining unchanged in the countryside for up to ten years. Problems occurred when the chemicals passed from animal to animal along the food chain with the result that all inhabitants of an environment could be affected.

The very toxic organochlorines - dieldrin, heptachlor and aldrin - were banned, including the dieldrin-based sheep-dips. Initially farmers were angry because they felt that the environmentalists had brought pressure to bear on the government. But residues of dieldrin were found in mutton fat in butchers' shops. As with birds of prey, these chemicals were also thought to cause serious problems to humans. Many birds of prey eggs were sterile, due to sub-lethal doses of these chemicals.

Although organochlorines, dissolved in soil water, did not necessarily affect, for example, a rose tree, or the greenflies which live on it, they were thought to have disastrous consequences for birds like the blue tit which eats vast quantities of greenfly. Birds of prey, which fed on large numbers of smaller birds, increased the concentrations in their own bodies. In some cases this killed the bird, and in other instances affected breeding success. Organochlorines taken into the body are stored in the fat without any known harmful effects. But at times when the fat is broken down for food, these chemicals are likely to be released with harmful effects. Although organochlorines have been replaced by other less lethal poisons, the use of these new substances continually adds to the world pollution problem. Not only have inhabited areas of the world suffered, but penguins in Antarctic wastes and marine fish in the ocean's depths, have all been affected and contaminated.

Pesticides were the first chemicals to be used, followed by herbicides which are still used on a wide scale, affecting Norfolk's flora. Herbicides which are used for killing weeds in crops, also drift on to adjoining hedges, heaths, woodlands and marshes, which also become polluted. Rapid decreases in the food plants of insects, which in turn are beneficial to the farmer, have given cause for concern.

The majority of the British Isles was originally covered with woodland, but today no natural woodland survives, with the exception of Wistman's Wood on Dartmoor. Forest clearance has left relatively few wooded areas. The reasons for this destruction are various: land was needed for housing, for cattle grazing, and for growing crops. Woodland was also felled to provide timber for fuel, houses and ships,

Kestrel - numbers increasing

and for numerous other purposes. Unlike some other habitats which are destroyed, woodlands have either not been able to regenerate or take much longer to do so. The intensive use of the cleared land prevents regeneration taking place. Many birds were originally woodland species, and although some have managed to adapt to new habitats, others have become extinct.

Areas of wetland and marshes, both inland and around the coast, have been reclaimed for farming and, as these habitats disappear, the plants and animals which live there are also likely to disappear. If habitat change was evolutionary - over a long period of time - the plants and animals would have a better chance of adapting to new conditions.

Other factors are responsible for changes in the countryside. One of these is more leisure time, coupled with an increase in the car-using fraternity. Few, if any, areas remain 'untouched' because few areas are inaccessible. The education of countryside users is therefore essential. But education does not touch everyone, and damage can be caused by people who have little concern for the countryside. Removal of rare plants, the careless starting of fires, and the dumping of rubbish, all have their own consequences. Bottles either 'dropped' at picnic spots or thrown away, often become death traps for large numbers of small mammals.

With an anticipated continuing increase in the population, changes in the countryside are likely to be greater. Humans are indiscriminate animals and it is only relatively recently that we have realised that future planning is an integral part of our well-being. Sprays used in the countryside might ultimately result in the death of man himself. In Norfolk - principally a farming county - the risks must be higher than in other more urbanised areas.

In the past, the farming community was encouraged to produce as much as possible. With large surpluses, this is no longer feasible, and set aside is now a reality. With this in mind, and the advice from the local Farming and Wildlife Advisory Group, farmers have an added impetus to consider conservation. The community at large also demands that wildlife conservation is the responsibility of us all, and ways are continually being investigated to pay farmers who manage their land in a way which is beneficial to wildlife. To do this effectively is costly, and can lead to a decrease in production for which farmers must be compensated.

With careful planning and fore-thought it is possible that more areas of the countryside can be saved for future generations to enjoy. It is impossible to imagine a world without wildlife, but at the present rate of change this could be a reality in the not-so-distant future - a world without birds, or other animals, without flowers or trees. But how many of us really care passionately enough to do anything about it?

Aconites piercing the snow

Photo: Nicolette Hallett

SPRING

SPRING IN NORFOLK

*First signs of life - Catkins - Coppicing - River plants - Nesting birds
on land and water - Butterflies - March hares*

'The hawthorn whitens, and the juicy groves
Put forth their buds unfolding by degrees
'Til the whole leafy forest stands displayed'

From 'The Seasons - Spring'   James Thompson

Just as the winter has been with us bringing its cold and bright and
mild and wet days, so spring will come, just as it has done from time
immemorial. During the winter some plants and animals have been
resting, waiting, perhaps impatiently, for the warmer days which
nature knows must lie ahead. But although life has not ceased, it is
true to say that during the last few months, things have been much
'quieter' in the county. Yet under the surface of the soil the early plants
are urging their heads upwards, to thrust the first tiny shoots above
the sombre, seemingly barren earth. Birds will be arriving as others
prepare to leave us. Animals are scuttling about looking for food.
The first leaves will soon appear on the trees and, unlike the drab
winter coat, each will be clothed in an infinite variety of greens,
interwoven and intermingled by nature as only she knows how. In fact
a whole new world is beginning to stir, awakening hopefully to a
bright, new year ahead.

Some life has already appeared; the yellow flower of the celandine
flaunts its shining head in the gardens and along the lanes. Swaying
gently in the breeze, too, are the white flowers of the wood anemone,
proudly proclaiming that spring is on the way. .All winter these resting
plants have been waiting, silently hidden beneath the brown,
inhospitable soil, all ready to burst into a frenzy of activity. And
around and above them, affording a kind of shelter and protection,
bushes and trees will soon be breaking forth in a show of splendid
greenery. Last summer each tree produced buds to be protected
through the winter by the scale leaves. Black ones covered the ash,
green ones the sycamore and birch and brown ones the beech. Oak

and horse chestnut buds were concealed by brown scales. The latter also had extra protection in the form of a sticky coat. Secreted within the horse-chestnut bud are many small leaves, each a miniature replica of the large horse-chestnut leaf. A wool-like covering within the buds affords protection for these tiny leaves. Their dormancy over, life has already begun to stir. As the leaf grows bigger the bud begins to swell, gradually pushing away

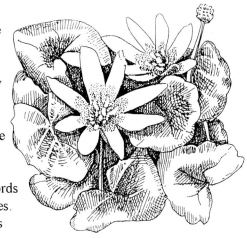

Celandine - heralds the spring

the scale leaves. Inside each bud is a short stalk from which the leaves arise. As this lengthens, the leaves are gradually pushed further apart, although the scale leaves remain attached to the stalk. They eventually fall, leaving a girdle scar on the twig. As the leaves emerge they are 'creased' where they have been folded inside the bud. Soon these creases disappear as the leaf grows in size.

One of the most fascinating sights of spring is the appearance of catkins of many kinds and shapes. The first are the long, slender golden male flowers of the hazel, which appear in February or March, and which country-dwellers call 'lambs tails'. The ripening catkins produce large amounts of pollen, and even the slightest breeze scatters clouds of the yellow 'dust' into the spring air. Each male catkin is composed of more than one hundred small flowers, which hang pendulum-fashion on a drooping stalk. When these catkins were formed on the tree last summer, they were very small, green and stiff. Here they remained relatively 'inactive' until spring when development gathered pace. Because the leaves only appear once the catkins have died away, they do not hinder the dispersal of the pollen. Shed by the male catkins it falls on to the female flowers, pollinating them, so that at least some familiar hazel nuts should develop by autumn. The female flowers are small and not easily discernible, but they are worth

Left: *Cowslips*

Photo: Gillian Beckett

Below: *Cuckoo pint in spring*

Photo: Ron Payne

seeking out because of their delightful crimson colour.

There are other catkin-bearing trees and shrubs which are also wind-pollinated, including the birch, oak, and poplars. But not all catkins rely on the wind for pollination. Some are dependent on insects, and, to encourage them, individual catkins are made up of a number of scales, each of which has a nectary producing a sweet-tasting liquid, supplying visiting insects with much-needed nourishment.

Hazel which grows in woods amongst the other broad-leaved trees produces less intense shade than other species like the oak and ash, enabling more plants to grow on the woodland floor. In days gone by, areas of woodland dominated the Norfolk landscape, and hazel grew in large numbers. The tree was useful to our ancestors who cut it back in the oak woods every ten years or so to allow the oak to grow to its full stature unimpeded by the hazel's branches. This was known as coppicing - which means to cut - and it is still practiced today in some areas. The long, straight hazel shoots which were removed during coppicing were used for thatching spars, fishing rods and walking sticks. Other wood was an important source of fuel.

Not only are the woods and fields showing signs of new life, but down by rivers and streams flowers are also beginning to appear. The yellow heads of both lesser celandine and marsh marigold - or kingcup - add gaiety to the tumbling rippling water. Where banks are drier, the tiny pink flowers of doves-foot cranesbill appear. According to legend they open to welcome the swallow, one of our favourite spring visitors.

Winter bird visitors, like fieldfares, redwings, knot, brent geese and goldeneye, are preparing to return to their own countries ready to breed. Our native bird species are joined by spring and summer migrants. Nest-building will soon begin and some birds have already started searching the countryside for a variety of materials. Moss, grass, leaves, twigs, straw and materials like wool are used. The degree of care taken when producing a nest varies from one species to another. Some birds spend a long time constructing a neat nest, but others throw together a few pieces of material - and there are even those which do not bother with a nest at all. Some birds never use the same nest twice; others come back to the same one season after season. Herons and rooks return to their tree-top haunts to repair and renovate last year's structures. Swallows attempt to find their old nests and set about repairing them. Sites chosen by different species are as varied as the construction and material used for the nest. The lapwing nests on the ground, and the house martin under the eaves of a house. Nestlings which are less mobile need a place which provides them with more security and also prevents them from falling out.

The wren is one of our smallest birds and the cock usually constructs several nests for his mate before she chooses the one where she eventually lays her eggs. She puts her stamp on it by lining one of the male's efforts with feathers.

Wren - diminutive songster

The collared dove is one species which has appeared on the scene relatively recently, the first British nesting birds being recorded at Cromer on the North Norfolk coast in 1955. After its first successes it has colonised many areas around the county and the country. In March both cock and hen birds help with building the flat nest, using stems, and twigs, often erected quite high in a tree It will also select a variety of other sites.

Many birds which depend on water for their food are also making preparations. Drake mallards (male) acquired their breeding plumage last autumn, and some ducks (female) begin building nests in February, laying their eggs later in the month, although nest-building and egg-laying may continue into May. As the light fades, flocks often leave their watery haunts for pastures new, where they feed on new shoots and the seeds of wild plants.

Two other birds associated with the water are the coot and moorhen. Both male and female coots take part in building a large nest in shallow water. Dried water plants are favoured, and although some nests are placed in and around water, other birds may build some distance away, perhaps in a bush or tree. The moorhen may nest in a shrub or tree some distance from the water. Both birds actively defend their territories. Male coots often fight fiercely in the water, and in these often violent attacks broken bones may result from the affray.

The large, graceful, grey heron has probably already begun to repair its nest or started building a new one high in the tree-tops, and some may already have laid. Communal nesters, they congregate in heronries high above the ground. The nest is a large platform-like structure and once built is repaired year after year. The female uses plant material together with twigs which the male brings to her.

The delicately dazzling and vibrant colours of our butterflies have been absent from the winter scene. British butterflies can be divided into two major groups. There are those which migrate, similar to some birds, and those which are resident. Included in the latter groups are small tortoiseshell, peacock and brimstone, which hibernate as adults, and these are possibly joined by some red admirals. In addition a further fifty-three species are considered resident, and other common butterflies in our region include the large (or cabbage) white and the red admiral. Of these the latter is a migrant, and the former a partial migrant. Resident large whites spend the winter as pupae and the population is supplemented by migrants from the continent. Some butterflies may begin their hibernatory sleep as early as August. In their winter refuges they hang in a state of 'suspended animation', as if they were dead. In her wisdom nature has ensured that the essential life processes of these insects are brought down to a level at which life still goes on, although it is barely discernible.

Spring would not be spring unless mention was made of 'mad March hares'. March marks one of the hare's main breeding seasons, although some may already have started in February and in many places breeding may continue until September. These 'crazy' antics are part of their courtship display. A pair may quietly be nibbling grass when, for no apparent reason, they take to their heels, kicking and jumping as they go. It has always been assumed that it was male fighting male, but recent research seems to suggest that if the female becomes 'bored' with the male's persistent attention during courtship, she will turn around and give him a few punches with her fore-paws. It is not unusual for several males to attempt to win the attention of a single female. Other hares may suddenly appear, pause with ears erect, and then dash off in pursuit of the 'free' female. After their frantic activities they quieten down and return to nibble the tender green shoots, only to take off perhaps a few minutes later in another wild chase. After mating, the hares go their separate ways.

There are other signs that spring is on the way; a new-found warmth gradually settles, virtually unnoticed, on the soft, brown earth, ready to change Norfolk's almost drab winter coat into a lively, exciting, galaxy of new-found colour and vitality.

Hares boxing

AMPHIBIANS AND REPTILES

Grass snakes and adders - Newts - Frogs and toads -
Cold-bloodedness

The British Isles has few reptiles and amphibians when compared with many other parts of the world and this is due to the position of our island. During the Ice Age most snakes were forced to move south. Although they returned to some of their former territories, as the ice continued to thaw the seas became full, separating the British Isles from the rest of Europe, cutting off any reptilian advance. The British Isles has six reptiles, of which three are snakes and three lizards, but not all occur in Norfolk.

Snakes are reptiles and Norfolk's best known inhabitant is probably the grass snake. These frequently occur near fresh water, where they swim to catch their food which includes frogs, toads, fish and newts. On land they take young birds, eggs and small mammals. Although grass snakes may reach a maximum length of four feet (120cm), on average females are no more than three feet (90cm) and males two feet (60cm). The largest snake in Britain, the grass snake often alarms people with its rapid hissing and constant head movements, but it seldom bites and even when it does, this is harmless. The reptile sometimes also releases a repugnant-smelling liquid.

When it is time for egg-laying - usually in June or July - the female chooses a place which provides a natural source of heat and compost heaps are often used. She lays some forty eggs and it will be three or four months before the young snakes emerge.

The other snake found in Norfolk is the adder - or viper - and its numbers are possibly on the decrease. Although venomous, its bite rarely kills. Smaller than the grass snake, an unusually large female may reach just over two feet (60cm). It is distinguished from its relative by a zigzag line down the length of its back, and a 'v'-shape on the head. Although the body colour varies widely from grey through white to yellow, browns and reds, the darker zigzag line always stands out. The most beautiful variation of adder is the black adder, which occurs from time to time.

Behaviour-wise the two species differ. When approached, the grass snake normally moves off rapidly into the nearest undergrowth, but not so the adder. It coils its body and waits for the 'next round', and remains in this position unless it is touched, when it is likely to defend itself by biting. The adder's prey is similar to the grass snake's and it feeds on nestlings, lizards and small mammals. Unlike the grass snake, the adder does not lay egg. Instead, the young develop inside the body of the female, hatching as soon as they are born, usually either in August or September. Again, it differs from the grass snake because it does not often swim, generally inhabiting drier areas.

Adder (top), Norfolk's only poisonous snake,
with the harmless grass snake

The most common of Norfolk's lizards is, perhaps not surprisingly, the common lizard. Like the adder, its young are born alive. It occurs in a wide variety of habitats, from country lanes to wooded areas. A 'wily' inquisitive animal, it always seems to be on the hunt for food. The common lizard has a wide-ranging diet which includes woodlice, insects, spiders and millipedes.

Norfolk is also home to the slowworm, Britain's only legless lizard. Inappropriately named, the creature is capable of quite a fair turn of speed, but it only generally shows this ability when disturbed. The Latin name is *Anguis fragilis* - *Anguis* means 'snake-like' and *fragilis* refers to the slowworm's ability to detach its tail.

Slowworm - neither slow nor a worm!

Having hibernated underground during the winter, slowworms may emerge towards the end of February or early in March if the weather is mild. The most secretive of the lizards, the slowworm is difficult to record, but it is estimated that the population is in decline because of the changing state of the Norfolk countryside.

They are found along hedgerow banks, on the edge of forest rides and in other undisturbed grassy areas. Slugs make up the bulk of the food, with the small white slug (*Agrolimax agrestis*) being the favourite. The lizard also takes worms, spiders and various insect larvae.

Newts are often mistaken for lizards, although they are amphibians. Three species occur in England: the palmate, the great crested and the smooth (or common).   The common newt, often called the smooth newt, is the most common species in Norfolk and the east of England. On average, the common newt is about four inches (10cms) in length, and dark green-brown in colour, broken by small darker-coloured blotches.   Underneath, the belly is generally reddy-orange, with a number of dark brown spots.   In some females these spots are not always present, and the underside is often lighter than in the male. Blackish dots break up an off-white throat.   Like frogs and toads, it needs water for breeding which usually takes place between April and June.   Because female newts do not move far from water their annual migrations are less noticeable than those toads.   Both weather and temperature have some effect on the earliest time at which breeding commences.

The legs of newts are not positioned directly under their bodies as they are, for example, in horses and cows, but are slightly to the side, and this gives the creature a splayed-out appearance when it walks. The limbs are similar in nature to those of man.   The two front feet have four toes each, and the rear five each.   There are no nails and the digits are rounded at the end.   Newts feed on insects, worms, waterfleas, snails and tadpoles.   There is a slit-like mouth, with teeth inside the jaw, which help to hold the prey.   The two eyes are positioned on the side of the head.

The great crested newt is a protected species, and for much of the year the difference between males and females is slight, although she may be slightly longer.   However, during courtship and breeding, there is a complete contrast.   The male produces his 'breeding' coat and a large showy crest appears along the back, which extends down the length of the tail.   The belly also takes on a more distinctive orange colour.   Although they live where the air is moist - perhaps under stones, in rotting logs and in amongst decaying vegetation - male newts wander far away from water.

All three species of newts lay their oval eggs singly in water and the female places these on the leaf of an underwater plant, which she then folds over.   Because the eggs are sticky, the folded leaf usually

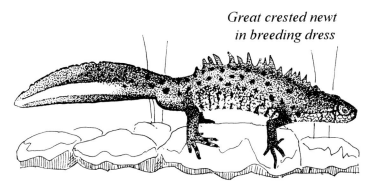

*Great crested newt
in breeding dress*

remains attached to the eggs. Once the eggs are laid, the female takes no further interest in her forthcoming offspring. After breeding, newts normally leave the water to spend the rest of the year on land.

When the eggs hatch, the young are similar in shape to the adults but without legs, and they have feathery gills. The front legs appear first, followed in 8 or 9 weeks by the back limbs. By this time the young are miniature replicas of their parents, but they still have their feathery gills. Small water animals provide the newts with a supply of food, and in turn the larvae of some pond dwellers, like great diving beetle larvae, will prey on them. The once-transparent skin becomes opaque, and by autumn the young newt, which is some three months old, has lost its gills. The animals leave the water to spend the winter hibernating on land. Any young which do not hatch until later, may spend the winter as larvae. It takes three to four years for the newts to reach maturity and become ready to breed.

Like newts, frogs and toads are amphibians - a word which means 'double life' - and water is vital for them, because they lay their eggs here. Frogs have no tails and a moist slimy skin, which must stay damp if they are to survive. Frogs have two methods of breathing. Oxygen is taken into the lungs from the air and through the skin. When a frog jumps into a pond it can sink to the bottom without drowning because it closes its nostrils, preventing water entering the lungs. Once oxygen in the lungs is used up, the animal breathes through the skin. Frogs are unable to stay under water indefinitely, and unless they come to the surface to breathe, they drown.

During the winter, frogs sleep buried in the mud at the bottom of

21

ponds and possibly ditches. In the past, when many farm ponds were filled in in the winter, large numbers of frogs were buried alive. In spring, frogs make their way from their winter quarters to their spawning ponds. When the eggs are laid - usually in the middle of the night - they are covered with a jelly. At first they sink to the bottom of the pond, but rise to the surface once the 'jelly' swells. How long it takes for the eggs to hatch depends on the weather, but in ideal conditions tadpoles may appear after ten days. and they hang on to the remains of their egg sacs, using up any remaining food. Gills take oxygen from the water. Gradual changes take place over the next few weeks until the young frogs resemble the adults. They leave the water, usually after a shower of rain, for a life on land. Although replicas of their parents, the froglets are only about half an inch (12mm) long, and it takes three years for them to reach maturity when they are ready to return to the water to breed. Despite the large numbers of eggs that are laid, few frogs reach maturity because, throughout their life, they provide food for a variety of other creatures.

The toad has a drier, warty skin, which produces a mild poison, causing some distress to its enemies. Toad eggs are laid in strings and wrapped around aquatic plants; when they hatch the tadpoles are similar to frogs. Toads return to the same areas of water year after year, often travelling considerable distances from their winter sites. Toad patrols are a common feature in some areas as humans collect them from roads to prevent large scale 'massacres'. The rescued toads are taken to their spawning grounds.

Of the three species of amphibians in Norfolk, the natterjack is the rarest. Although similar to the common toad, the vast majority have a distinguishing yellow stripe down the middle of the back. Common toads crawl about; natterjacks can often be seen running. In Norfolk natterjacks are found in sandy places, with well established colonies at Holme Dunes Nature Reserve. They can often be found where there are small burrows or may take to living under debris. They are 'out of their depth' in Britain, because of the unpredictable nature of the climate, and they are happier in warmer parts of the world. Most are found in the lowland heaths of Britain and in some of Norfolk's sandy areas - where it is the warmer. Sandy soil is important because the natterjack needs a loose substrate in which to burrow. When it comes to breeding, the natterjack prefers shallow ponds, and this often poses a problem because such areas may dry out in summer before the natterjack tadpoles are fully grown. But it seems that as long as they are able to breed successfully every two to three years they can survive.

Natterjack toad -
a lover of sandy soils

Reptiles and amphibians are cold-blooded animals and unlike humans, whose temperature remains fairly constant, the temperature of these animals is similar to their surroundings. By going into hibernation they overcome the cold conditions and the rate of breathing falls to such a low level that they could be taken for dead. Frogs bury themselves under moist mud because, although virtually 'dead', the skin must remain moist to enable them to breathe. Snakes, lizards and toads do not need a damp environment, and all three groups choose drier areas, usually where there are leaves. Apart from the cold, which would freeze their body fluids, the food supply also disappears.

SPRING FLOWERS IN DANGER

Norfolk's horticultural heritage - Herbicide spraying -
'Country-lovers' and their cars - Cornfield plants - Norfolk's orchids
Cuckoo pint and cowslips

Norfolk is rich in wild flowers, and much botanical work has been carried out in the county. It is probably true to say that Norfolk's flora has been studied as much, if not more, than most other counties in the British Isles, since many eminent botanists have lived and worked in the county. The first written account of Britain's plant life appeared in 'Herbal', written by William Turner, and published in 1551. But much earlier than this a great deal of interest had already been shown by inquisitive botanists in the plants which were found in Norfolk.

During the reign of Edward III, Flemish weavers came to Norfolk, and within a relatively short time they had established a very profitable weaving industry in towns like Worstead, Aylsham and North Walsham. During the second half of the sixteenth century a substantial Stranger community of clothworkers from the Low Countries was established in Norwich, which was then held by some to be the second largest city in the British Isles. When new 'colonists' arrive they often bring native plants with them, and this was true of the Flemish weavers. Gardening was one of the hobbies of these earlier settlers, and the local inhabitants soon 'caught on'. In those early days gardens, as well as cities, were usually walled in. To supplement the plants which they found in their gardens, the new arrivals probably went outside the city boundaries to collect other species or had seeds sent to them from Holland. Norwich was one of the first places in the country to have its own horticultural society. From about 1500 onwards, when the printed word became more widely available, until the present day, many interesting books have been written and published about Norfolk's botany.

But many of these wild plants which the early botanists 'discovered' have either vanished or are now very rare. This has happened - and continues to happen - because a variety of factors combine to affect the Norfolk countryside, like so many other areas of the British Isles. And the question often asked is 'Where have all our wild flowers gone?'. It is true that there are still large numbers of wild flowers, but many have either become extinct or endangered in the county.

There are various reasons why Norfolk's diversity of wild plants continues to diminish. Changes in agriculture, leisure, and in the countryside in general, account for this. A Private Members Bill has gone a long way to protecting our wild plants, particularly the rare species, and it is now an offence to pull up any wild plant without permission from the owner of the land. However, activities like the re-designation of Sites of Special Scientific Interest (SSSI's) caused the demise of some plants because some areas were destroyed before they could be re-notified.

Factors responsible for the decline of wild plants, so much a part of Britain's heritage, include the regular spraying of crops with herbicides. These kill what naturalists call 'wild plants', and which farmers call 'weeds'. No one is suggesting that crops should not be treated, but unfortunately the effect does not stop here. Even on a still, virtually wind-free day, the drift from spraying a field can be considerable. Large areas of vegetation, not associated with the sprayed field, may be affected, including heaths, woodlands, roadside verges and hedgerows.

Although spraying may be one of the most noticeable effects, there are others. Woods are cut down and plants, which are only capable of living in deep shade conditions afforded by the trees, are no longer able to survive. There is much wildlife in the countryside which is able to adapt to new conditions and plants are amongst these. But sudden changes provide little time for this to take place. Perhaps one of the earliest disasters as far as wild plants were concerned was the arrival of the motor car. Prior to this, at least some parts of the countryside were relatively safe from the onslaught of large numbers of people, but this is no longer true. With the coming of the car few areas can now be classed as inaccessible, and when people first came to such places they often dug up wild plants which they had not seen before to take back for their gardens. In many cases these plants did not survive, and people returned time and time again to tear up replacements. Within a relatively short time plants which were locally common were in danger because of the thoughtlessness of so-called country-lovers. This still happens to a lesser extent in spite of legislation to protect wild plants.

During and after the last World War, every inch of the countryside was needed for growing crops. To provide this additional land, bogs and marshes were drained. It was these 'damp' habitats which harboured many rare species, and with drainage completed a unique flora became extinct virtually overnight. To increase the area of agricultural land, hedgerows were also removed at a remarkable rate. Where hedgerows remained, flowers which grew there were killed by chemical sprays. When these were banned, hedgerows were 'attacked' by mechanical cutters. Towards the end of the 1990's, with the decreasing need for such intensive farming, agricultural land is being 'set aside', which has led to many roadside verges and strips around the field margins being left for wildlife to colonise. Many farmers are planting wild flowers in these areas.

Today's corn is threshed much more efficiently than previously, and the seeds of what the farmer calls weeds are meticulously removed. Not so many years ago the cornfields of Norfolk were alive with wild flowers from late spring to autumn. The scarlet patchwork of thousands of poppies swaying in a summer breeze, is a sight which is merely a memory to many people. But fields of poppies do appear from time to time because the seeds can remain dormant in the soil for long periods. This was especially noticeable after the first year of set aside when fields were once again ablaze with the fiery heads of thousands of these distinctive flowers Herbicides have almost eradicated once common cornfield plants like corn cockle and corn marigold, although both are being planted in gardens. Cornflowers, with their scintillating blue blooms, have also disappeared. A hundred or so years ago a writer commented that plants, like the cornflower and the poppy were 'troubling the cornfields with their destroying beauty'. Where fields were not glowing with the heads of red poppies, they were often clothed with the golden flower heads of charlock. These cornfield plants had been growing in the fields for centuries, but were destroyed by poisonous sprays in a very short time.

*Corn cockle -
a disappearing species*

In some areas roadside verges have been set aside as reserves Posts bearing the letters 'NR' are used to show that the area has been 'left to nature'. In many places long lengths of verge are left and the vegetation is only cut in dangerous areas, like corners and at bends. With excessive rain, hedgerow plants, including the umbrella-shaped heads of species such as hogweed and cow parsley, become rampant Dandelions flourish along the roadside edges and grasses grow quickly.

Orchids are some of Norfolk's rarest species, and world-wide they form a very large family of perennial plants. The most majestic grow in tropical countries, but some fifty species occur in the British Isles. Of these, about thirty are thought to occur in Norfolk. Because of their diversity of form - some are beautiful, others are bizarre - they were picked by indiscriminate 'country-lovers'. There was a great deal of misunderstanding about these, and some people believed that removing the flower improved the plant. Although this is true for certain garden plants, it does *not* apply to orchids. Other people thought that uprooting a plant and taking it to their cottage gardens would do the orchid a 'favour'. Unfortunately the majority of orchids die if removed in this way, because their survival depends on a relationship with a fungus in the soil in which they grow.

Many species of orchids take a long time to grow and produce seeds. Even 'fast flowering' orchids may take between five and seven years to produce a bloom. In others, like the lady's slipper orchid, it is sixteen years before a flower appears. Of all Norfolk's orchids one, the rare bird's nest orchid, has very little chlorophyll (green colouring) which is necessary so that the plant can utilise sunlight to make food Its survival depends on the damp humus on which it lives, found on calcareous - chalk and limestone - soils beneath the shade of towering

27

woodland beech trees. An association is formed with a fungus in the soil, which enables the orchid to obtain its food supply. It is from the rooting system, and not the flowers, that the bird's nest orchid gets its name. These underground stems send up large numbers of short roots, which all grow at different angles from the main root. This root arrangement looks like a bird's nest.

Other rare orchids also grow in Norfolk, including the fly orchid. Seed production occurs once pollination has taken place, which is achieved by one of nature's clever 'tricks'. Small burrowing wasps mistake the shape of the flower for their own kind. They collect pollen from one orchid, and as they push into another flower pollination takes place. There are several other rare species including broad-leaved helleborine, autumn lady's tresses, bog orchid, frog orchid and greater butterfly orchid. In addition a number of other orchids, like the man orchid, have become extinct.

The fen orchid was suited to past conditions in Norfolk, and where fen farmers grew sedge, this was an ideal habitat for the fen orchid which grew in abundance. The sedge protected the orchid without smothering it. When sedge cutting stopped, the plant grew above the fen orchid stifling it, resulting in its extinction. The orchid also occurred when peat was removed, an activity which resulted in the formation of the Broads. At some stage in peat removal, temporary fen conditions appeared leaving holes which gradually filled with water. The fen orchid colonised these areas growing without interference, but as these evolved, the fen orchid was unable to tolerate the new conditions and died out in many places, although it is still recorded at a number of sites.

One of the strangest, and yet intriguing, things about orchids is that they appear, disappear, and then reappear again for no apparent reason, although it may be connected with pollination - or lack of it. In these plants pollination is a 'chancy' business and it has been suggested that the insects responsible for this activity are as rare as the orchids themselves. Artificial pollination has been successfully carried out at a number of sites.

If orchids are considered interesting then the cuckoo pint, variously known as 'lords and ladies', 'wild arum' and 'jack in the pulpit', is one of the strangest plants to be found in Norfolk. The central purple coloured part, known as the spadix, gives off a rather unusual, and to some people, obnoxious smell, not unlike decaying flesh. The flowers occur at the base of the spadix and flies, known as owl midges, attracted by the smell which the wild arum produces, slide down the slippery slope passing through a ring of hairs where male and female flowers occur. These hairs trap the flies, which carry pollen picked up from another wild arum plant; they drop this on to the female flowers. The hairs eventually wither and the flies escape, picking up pollen from the male flowers on the way out. When they find another wild arum flower they repeat the activity, and this 'strange' method ensures that seeds are produced.

One of the delights of spring in the past was to be able to take home the first bunch of cowslips from the local meadow. Thirty years ago they grew in many meadows in Norfolk, but they have become rare in the county. Pasture land has either been ploughed up or sprayed to improve the quality of the feed for cattle, eliminating the cowslip. On a happier note, with an increasing acreage being set aside, perhaps the cowslip will return. The oxlip, a relative of the cowslip, has managed to survive due to careful management. It is one of Norfolk's rare plants, and occurs in only one area, but fortunately the area where it grows, has been made into a nature reserve. Many young plants have grown around the main colony, which appears to be spreading. The oxlip is a member of the primrose family, but unlike other members, prefers moist areas. The Norfolk colony grows in the open on an area of peat fen, a unique habitat as far as this plant is concerned.

Left: *Small tortoiseshell
butterfly on
hemp agrimony*

Photo: Roger Tidman

Below: *Southern marsh
orchids*

Photo: Martin Smith

BUTTERFLIES

Cabbage white - Red admiral - Swallowtail

Having spent the winter asleep, the resting butterflies are stirred by the first warm days of spring and gradually they 'come to life'. The first butterflies of spring are the brimstones, small tortoiseshells and peacocks. In mild conditions it is not unusual for some small tortoiseshells to make their debut in winter, although many will perish. It is more likely that many will 'come to life' as early as February. Having slept for many months these relatively fragile insects have been drained of their energy reserves and their first task is to seek out any nectar-producing flowers to give them a burst of energy before they are ready to mate.

As the weather warms up more and more butterflies are enticed out. Some of these will have over-wintered as adults, but many more as pupae (chrysalides), larvae or eggs, in what can sometimes be harsh conditions. Why nature has chosen these different stages no one is certain, and the number which come through this difficult period varies from species to species. Just as birds migrate, so do some butterflies and the first to reach Norfolk from across the Channel are the large or cabbage whites. In some years there will be large swarms; in other seasons quite small numbers make landfall. The abundance or scarcity of the large white in the countryside is not due solely to the success of Britain's over-wintering pupae, but to the contribution made by visitors from afar. Having crossed the North Sea, the butterflies seek a place for the night. They will move from one plant to another until they alight on a suitable resting place. Because of their pale colour the butterflies must find light-coloured vegetation to be effectively camouflaged, otherwise they would stand out against a dark background and probably be quickly picked off by some hungry predator.

Having settled in, the butterflies top up their energy reserves before they mate. This activity over, the female is ready to lay her eggs in May or June. She seeks suitable plants belonging to the cabbage family - from which the large white gets its alternative name. Each

31

female produces between forty and a hundred skittle-shaped eggs at one sitting. Around fourteen seconds is needed for each egg to be laid. If the weather is suitable another egg-laying session may take place in August or September. The weather also affects the hatching time. When conditions are favourable the caterpillars emerge after four or five days. In damp weather it may take up to sixteen or seventeen days. This is one of nature's safeguards. Caterpillars which emerge in unsuitable conditions usually die either from the cold or damp.

Once out of the egg there is an urgency to feed, and the first nourishment comes from the discarded egg case which is devoured. Many leaves on which the eggs are laid are slippery and the newly-emerged caterpillar spins a layer of silk so that it can move about more easily.

Feeding continues for around thirty days. All the caterpillars on a particular leaf feed and rest in unison and during these interludes they all come together. When they resume feeding they move apart.

The large white caterpillar is grey-green and is usually darker on the back, with yellow stripes down each side of the body. During the caterpillar stage the skin is shed four times. The only way that the caterpillar can grow is to discard the old skin once its body has filled it. It spins a silk 'mat' and rests without feeding. By twisting and wriggling, the skin starts to split behind the head, enabling the caterpillar to poke this part

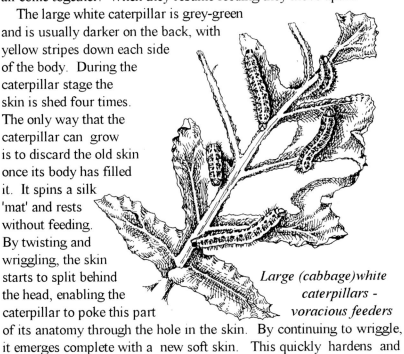

Large (cabbage)white caterpillars - voracious feeders

of its anatomy through the hole in the skin. By continuing to wriggle, it emerges complete with a new soft skin. This quickly hardens and the insect is ready to feed again - and three more moults are necessary before the caterpillar is fully grown and ready to pupate.

Once fully grown the caterpillar stops feeding and seeks a quiet and sheltered spot where it spins a silk pad and girdle. It attaches the latter to a suitable support and then rests with the end of its body on the silken pad. Another skin change leads to the pupal stage. If the caterpillar is from the early (spring) brood, the adult will emerge in about fourteen days. Caterpillars from the autumn hatch will remain as pupae for the next eight months. The transformation from caterpillar to adult butterfly is one of nature's miracles. The reorganisation of the tissues provides a spectacular butterfly from a dowdy caterpillar. As emergence time arrives the skin splits and the butterfly is released from the pupa. Helpless at this stage with wet, weak wings, it clings tenaciously to a twig. Liquid, which is pumped into the wings, fills them and they are soon fully developed. Once 'full' the butterfly forces its wings straight out, stopping the flow. The adult remains stationary for as short a time as possible to prevent the attraction of preying animals.

A few quick movements of the wings, and the butterfly flies off. The next 24-25 days are spent on the wing, with frequent pauses for nectar - imbibing sessions from flowers. Mating takes place and eggs are laid.

The wings of the male large white are a palish creamy-white, broken by either brown or black markings on the upper surface. The female contrasts with the male

Adult male large (cabbage) white

by having two round spots near the centre of the forewings. The hind wings are more yellow than those of the male. Although possessing compound eyes, sight appears to be limited and food and egg-laying plants are detected by smell.

Of the many common butterflies found in the county, the red admiral is one of the largest and most attractive. There is still some disagreement about the butterfly's ability to hibernate successfully during British winters. Unlike the small tortoiseshells, all of which seem to appear at the same time, mate, lay their eggs and produce succeeding generations, the red admiral's activities are more haphazard. The first sporadic sightings of the butterfly may take place as early as February in warm weather, but numbers do not increase until about May, with the insects reaching their highest population in late summer an early autumn. Although the butterfly hibernates in the adult stage, it is thought that only a small number survive. No one has explained why the red admiral perishes during our winters, when other species, like the brimstone, small tortoiseshell and peacock, make it through the winter, especially when red admirals survive in colder parts of Europe. Without doubt some make it across the North Sea, but how many successfully complete the journey is not known, and what happens to those which do not attempt the journey is something of a mystery.

As a migratory species the red admiral population is totally dependent on the numbers which reach the British Isles from the Mediterranean. Once these migrants have reached our shores they can be seen on the wing in a variety of habitats. They frequent woods and gardens, and appear almost anywhere in the countryside. Since numbers vary from one season to the next, the previous year's weather conditions must have an effect. Warm, dry summers boost the red admiral population and increase the chances of those going into hibernation. The summer of 1989 was a good year for the butterfly and F. W. Frohawk, one of the best-known naturalists at the turn of the century, recorded good populations in 1904.

The arrivals lay eggs as soon as they make landfall and then tend to move northwards. Later, triggered by some mechanism as yet not understood, the butterfly returns to the south, where large numbers congregate on the south coast.

Of the more common insects found in the county the red admiral is undoubtedly one of the brightest. During the summer and early

Above: Red admiral feeding off rotten apple

Photo: Roger Tidman

Photo: Roger Tidman

Below: Swallowtail on ragged robin

autumn it visits orchards and takes the juice from fallen apples. Autumn flowers, like michaelmas daisies and buddleia, prove a magnet for the red admiral, which, together with the small tortoiseshell and peacock, often completely cover the flower heads on warm sunny days. Here they imbibe the nutritious nectar to provide valuable energy either for migrating or to see them through the sleeping months which follow.

The success of these three butterflies is undoubtedly due in some part to their food plant. They need nothing more sophisticated than stinging nettles - and yet even these are not as common as they used to be, because they are sprayed or uprooted to keep the countryside free from weeds.

Although both male and female butterflies have similar markings, the female is usually easy to spot when she is egg-laying. The initial impression might be of a bumbling insect uncertain of her duties. But this is far from the case; she is obviously searching out the best food for the caterpillars once they hatch from the eggs she is laying.

Once she is satisfied, the female places a single egg on the growing leaf tip and, having left it, goes off to repeat the activity. Although stinging nettles are the major food plant, eggs are also laid on annual nettle and pellitory-of-the-wall. Once the caterpillars hatch, in about a week, they lead solitary lives. First the insect makes a tent by pulling together the edges of the leaves, which it binds with silk. The larva feeds inside its silken home without too much trouble from predators, although many fall victim to parasitic flies.

The caterpillar may be black or green, but both variations have yellow stripes down the side. After several moults and for about 28 days, depending on the weather, it continues to feed inside a new tent in what Moses Harris, an 18th century naturalist, called 'places of security'. With succeeding moults the tents change shape and are easy to spot on the food plant. It is not unusual for the insect to leave the tent for the last few days of feeding before pupation takes place. The new tent which it creates here is different. Having found a suitable fresh shoot, it bites almost through this. The tip falls and lands on the surrounding leaves. Now the caterpillar gets to work and makes silk for a cocoon which encloses all the leaves. Pupation may take place here if there is enough room. But more often than not, the caterpillar crawls away and forms a new tent by joining two or three leaves together.

The swallowtail is probably the rarest of Norfolk's butterflies. A native of the Broadland and fens of East Anglia, its numbers have declined dramatically and the Broadland colonies are the only ones to survive. But from past records it is known to have occurred in other parts of Southern Britain. The draining of its original haunts, coupled with intensive farming, has decreased its numbers to levels which give constant cause for concern.

In its Broadland home the food plant of the swallowtail - milk parsley - was often smothered by more rampant species like rushes, and when management was either non-existent or limited, swallowtail numbers declined to danger level in the mid-1970's. However a greater interest in the butterfly, together with a better understanding of management techniques and the insect's requirements, gave rise to an increase in the number of milk parsley plants, resulting in a growing population which is relatively healthy in the mid-1990's.

The swallowtail is undoubtedly one of the most distinctive and attractive of the British butterflies. With a heavy body, the adult has to keep the wings in motion to remain suspended above plants, as its long proboscis sucks up the nectar from a marshland flower. Feeding is usually confined to morning and late afternoon. During the breeding season, having taken enough nourishment, the males spend their day waiting for the arrival of the females. Prior to mating, there is an interesting courtship ritual. As the pairs come into contact with each other, they gently hover in any breeze, and then fly high above the countryside, before they plunge earthwards to land on the plants where mating takes place. The pair may remain on the same plant for several hours until she leaves in search of food plants to lay her eggs. The female appears choosy about the plants where she deposits the eggs, and she flies a few centimetres above a number before selecting one for her eggs.

Two weeks later the eggs hatch, and the caterpillars resemble bird-droppings until they reach their third moult. But in spite of this camouflage more than half the population is taken by spiders in these early stages. Later, birds replace the spiders as the main predators, with species like reed buntings, bearded tits and sedge warblers consuming large numbers, perhaps as many as sixty six percent of the remaining population.

Why the caterpillars are prone to such high levels of predation is a mystery, because they have a bright, fleshy, forked, orange-coloured organ on the front segment of the body. Called the osmetrium, it produces an unpleasant liquid which ought to keep enemies at bay! But this does not seem to deter would-be predators and the distinctive striped body and habit of sitting well up on the plant literally advertises the caterpillar's presence. Feeding continues for a month, after which the swallowtail caterpillar leaves the milk parsley to pupate on the stem of some reedbed plant.

Dog roses

SUMMER

And to itself the subtle air
Such sovereignty assumes,
That it receiv'd too large a share
From nature's rich perfumes.

From 'A Summer's Evening' by Michael Drayton

BIRDS OF THE COAST

Shoreline birds, oystercatchers and terns - Birds of the cliffs,
fulmars and sand martins - The gull family - Waders

Norfolk is fortunate in having a long, interesting and varied coastline, providing habitats for many bird species not found in landlocked counties - or even in inland areas of our own county.

Coastal birds can be divided into two broad groups according to where and how they nest. There are those which have their nests close to the edge of the sea so that when there are very high tides, either eggs or young - or even both - are destroyed. Birds which nest on the ground - on shingle ridge and sandy dune - also fall into this category. The second group includes those birds which prefer a somewhat safer position on the cliffs which border the sea in some areas of Norfolk.

The oystercatcher, a common bird of coastal habitats, has been referred to by more than one naturalist as 'the pied piper of the shoreline'. It is one of many species of birds to be found at places like Blakeney Point. It has a shrill, nervous cry which sounds like 'kleep-kleep', but when disturbed its call turns to an irritated 'pic-pic'.

Before they mate, groups of oystercatchers congregate for a piping performance. They form a circle and then run around with their beaks close to the ground uttering their distinctive piping call. Like many coastal dwellers, the oystercatcher does not bother with the sort of nest favoured by common terrestrial species like blackbirds and thrushes. Both birds make a number of scrapes in the sand, the female selecting the one where she will lay her eggs. The lining consists of 'debris' including bits of shells, maybe the odd pebble and perhaps some plant material. Some birds desert the shore and lay in fields close by.

The oystercatcher relies on camouflage to protect its eggs from predators. This is so effective that it is easy to walk on to a nest - except that the birds are always on duty, and will utter their 'kleepee' alarm call; other nesting oystercatchers will then help repel intruders.

Once the eggs hatch, the birds bring food to the chicks, an unusual feature of ground-nesting seashore birds, but this is because of the

40

specialised nature of the diet. If danger threatens, the chicks freeze on the spot and rely on the parents and neighbouring oystercatchers to protect them.

In bulk the oystercatcher is about the same size as a pigeon, but less drably coloured. Its plumage is a dazzling contrast of black and white, with a long stout orange bill, the tip of which has evolved to deal with its bivalve food - cockles and mussels. Chisel-like at the end, the bird uses this bill to hammer its prey against a suitable object like a stone. The bird may alternate this technique with stalking until it finds an open shell when it will pull the creature out.

The name 'oystercatcher' suggests that the birds feed on these marine creatures. But this is misleading although, strangely, it is also known by the same name in many continental countries. It does not feed on oysters, because these are usually found in deep water; instead it takes cockles and mussels. The number of oystercatchers found in a particular area at any one time is governed by the food supply. Although the bird wades into the sea for food, it never dives. It can often be seen flying low over sandbanks and channels along the coast. In winter the numbers increase as birds from Iceland, the Faeroes and Norway join the 40,000 or so birds which breed in Britain.

If the oystercatcher is the most striking bird of the seashore, there are many others which should not be overlooked. The common tern is one of these. It sets off from its winter quarters on the west coast of Africa in March and makes its way to the Norfolk coast, arriving in April and staying until October when it makes the long and arduous journey back to its African winter home. Although common terns are the most frequently seen, others, including the little tern and Arctic tern, also breed along the Norfolk coast. Blakeney Point has the distinction of having the largest breeding colony of little terns in the British Isles. The common tern is a graceful bird and often called the 'sea swallow' because of its prominently forked tail, seen clearly when the bird is in flight. Terns nest in colonies on the shingle ridges close to the sea, and they scrape a hole in sand or shingle, or sometimes in clumps of grass or rushes. Most nests are not lined, but those which are, are simply furnished with nothing more than a pebble, fragment of shell, bit of wood or some strands of dry grass to complete the interior

Above: *Common terns*

Photo: Martin Smith

Photo: Martin Smith

Below: *Oysterctacher nesting*

- a job undertaken by the female. The female lays two or three eggs which vary in colour. Some have a bluish-white ground colour; others may be green or buff. There are always brown blotches which break up the base-colour, effectively camouflaging the eggs.

Common as well as Arctic terns are aggressive during the breeding season and they will go to great lengths to defend their territories. Any intruder is dive-bombed with a raucous cry of 'keeyah', and it is not unusual for the sharp bill to draw blood from a head where it makes contact!

Difficult to distinguish apart, Arctic terns often nest in amongst the common tern colonies. The most distinguishing feature is that the common tern has a dark patch on the front of the wing and an orange-red bill, compared to a blood-red bill in the Arctic species. The white underpart of both birds is in contrast to the black crown.

The largest of the breeding terns is the Sandwich tern, and its legs and beak are black, the latter having a yellow tip. The number of breeding Sandwich terns has increased in recent years because of much better protection in their summer coastal homes. At one time, another species, the little tern, was the rarest of the terns breeding in Britain, and although its numbers are still relatively small, the protection it has been given has paid dividends and its numbers tend to increase annually. It has a black-tipped yellow beak and white forehead which, apart from its size, are the most noticeable features.

Travelling eastwards, the cliffs at Old Hunstanton give way to flat expanses of sand and shingle, which are dominated by rising ridges of shifting sand dunes - and then revert to cliffs again at Weybourne. It is here that fulmars find some of their nesting sites and, using the many thermals which are a feature of this part of the Norfolk coast, these seabirds glide effortlessly and almost unceasingly. Over the years the nesting sites have changed and more fulmars are found at Hunstanton where there is a thriving colony. These aerial acrobats are best seen from the cliff tops as they rise and fall above the waves. Fulmars have brilliant snow-white heads, contrasting sharply with the back which varies from an almost nondescript grey to a more definite sooty black. Unlike the terns, which are migrants and return to their winter homes, fulmars spend much of the winter along the coast.

Here they stay, gliding out to sea on calm days in search of food. The number of birds steadily increases until a climax is reached in the first weeks of April, when the fulmars select their nesting sites. A fascinating courtship display takes place in the same month, followed by egg-laying towards the end of May. The natural nest consists of a depression in the rock. Occasionally the nest is lined with the odd pebble. A single egg is laid and if this is 'lost' it is not replaced. Once the eggs are laid, the fulmar moults. Both parents take responsibility for their incubatory duties, and also for feeding the chick, which remains on the rocky ledge protected by the parents during the early days. As the young fulmar grows, the adults leave it and spend more time out at sea fishing, but continue to feed the young bird for around 53 days, after which it is ready to leave. Having fledged, the young birds are ready to fly. Some will take off straight out to sea, but the less 'agile' birds fall down the cliffs when they first exercise their wings.

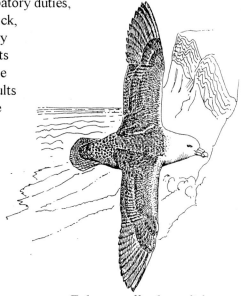

Fulmar - effortless glider

Even in atrocious conditions fulmars spend many months out to sea - and this is especially true of immature birds which are truly oceanic. Unlike some bird species, where numbers are on the decline, the fulmar has made remarkable headway in the last eighty or so years. In the 1880's the only records were from St Kilda in the Hebrides. The birds first appeared in Norfolk in 1939, when they were seen on the cliffs at East Runton. Although the population increased, the birds did not breed until some time after the initial invasion. In spite of the fact that large numbers of fulmars are recorded along the cliffs of

44

Norfolk, few of these are thought to nest. The main breeding ground of the fulmar has moved from the cliffs between Mundesley and Weybourne to similar sites at Hunstanton. Of all the petrels, fulmars live for the longest time, with many reports of 40 year old birds.

These same cliffs provide suitable, although at times frustrating, sites for sand martins. The cliffs are suitable because they are composed of soft rock, which enables the birds to excavate their nests. The frustration occurs because of almost ceaseless coastal erosion. The nest, which is generally 2-3 feet (60-90m) into the cliff, is often destroyed. The nest excavated, the martins search the drift line for suitable lining material. Seaweed is most common, supplemented by grass from the cliffs themselves. At Hunstanton, the cliffs are also home to the house martin. Here, and nowhere else along the coast of the British Isles, there is a breeding colony of house martins in the cliffs. The colony is an old-established one, and nesting birds were first recorded in the late 1880's. This is the kind of site where they must have nested for millions of years before houses existed.

Several species of gulls are found in a variety of areas along the coast, as well as in innumerable inland habitats. The black-headed gull, one of the smaller species, changes its plumage in the breeding season. Normally pale grey, both on the wings and back, the under surface of the wing is white, and white feathers continue along the leading edges, contrasting sharply with the black wing tips. The bright red legs and bill add to the bird's distinguishing features. At breeding time the head takes on a chocolate brown appearance, but these feathers are moulted once mating is over, although a dark conspicuous spot can still be seen behind the eye

Slightly larger than the black-headed gull, the common species has more white plumage, with both underparts and head also showing this colour, and although the wing tips are black, they have distinguishing white spots. The bill and legs are yellow-green.

To the uninitiated all gulls are referred to as 'seagulls', but it is the kittiwake which truly deserves this title because it spends most of its time out at sea - especially in winter. Similar to the common gull in shape and colouring, it lacks the white spot on the wing tips. The bill is a deeper and more distinct yellow than its common relative. Unlike black-headed and common gulls - which prefer ground-nesting sites - the kittiwake is gregarious and nests in colonies on cliffs.

Two other large members of the family are the great black-backed gull and the herring gull. The former is a large bird with black wings, back and head. Similar in size, the lesser black-backed gull is lighter in colour when compared to the great black-backed species.

Although adult gulls may be relatively easy to recognise, immature birds prove something of a problem, because they all appear to have similar colouring, with a drab, mottled, brownish-grey plumage. These juvenile feathers are retained until the birds are two or three years old.

Of the smaller wading birds, one of the most noticeable is the ringed plover. A small, plump bird, it is easily identified by a wide black band, which breaks up the white chest, and the white collar. Local people call it the 'stone runner'. Like the terns, it nests by making a shallow scrape in the ground which is sometimes lined with bits of shell or small stones. When the young chick hatches it is difficult to see, and a trained eye is needed to follow the well-camouflaged 'ball of fluff' as it dashes across the sand. If disturbed, the chick 'freezes' on the spot, only continuing its activities once danger has passed. To distract attention from their offspring, the parents put on a broken wing display.

The bobbing head of the redshank, and its conspicuous bright legs, are good identification points. Unlike many other coastal dwellers, the redshank makes its nest in long grass. Like the ringed plover, the chicks start to run around almost as soon as they are hatched.

The ground-nesting ringed plover

LIFE IN PONDS

Variety of pond life - Mayflies and caddisflies - Great diving beetle -
Backswimmers, water scorpions and water spiders -
Pond snails - Mallards

Many of the insects which are around during the summer months,
like the dragonflies, damselflies and caddisflies, live in ponds, in a
form completely different from the adults. But the larvae of these
insects are not the only inhabitants of the world of the pond; many
other animals live there too.

Frogs are often the largest occupants of the pond, apart from some
fish and the birds which swim on the surface and dive for their food.
But there is a variety of life which carries on with its daily 'chores'
unseen by most of us. These pond creatures move around in different
ways; some swim, others crawl. There are those which
spend their existence on the bottom of a pond; others
may inhabit an area close to the surface.
Animals related to seashore creatures - the
crabs and shrimps - including water fleas and
freshwater shrimps, roam around in the pond.
There are also several species of snails, as
well as worms. These animals are true pond
creatures; their life begins and ends in the
pond. But, there are other animals which
may spend only part of their life in the
water, including a variety of insects.
In the case of some pond dwellers the
life history may be long, in others
it is over rapidly.

The shortest-lived insects which begin
life in the water are the mayflies. The poet
Shelley said that they 'gathered into death
without a dawn'. The young stage, the
nymph, may spend
more than two years   *Freshwater shrimp -*
in the water.   *food for many pond dwellers*

47

Some species are free swimming, but others bury themselves in the mud at the bottom of the pond. Nymphs are easy to identify because they have three long tail filaments. A 'creature' emerges from the nymphal case, but this is not the adult, and it must moult before the adult emerges.

The airborne adult has a short life and, depending on the species, may survive for only a few hours. Male mayflies swarm over, or close to, water from May to September when they perform their dancing 'nuptial' flights. The females join them and after mating the eggs are laid in water. In some species the female pushes the abdomen just below the surface. In others the eggs are dropped into the water or the mayflies may go into the water to lay their eggs.

The caddisfly larva is a fascinating pond-dweller. Some build cases in which to live, others make use of a hollow stem, whilst yet more are free swimming. The material used for case-building varies from species to species, although all make a basic silken case. Some caddisfly larvae then give the silken case an extra layer which may be made from vegetation, small stones or grains of sand. The front legs of the case-carriers protrude, enabling the insect to crawl. Before pupation the insects move to shallower water at the edge of the pond, and bury themselves in the mud. They close the end of the case almost completely, except for a small hole which allows fresh water to enter and escape. Ready to emerge, the pupa pushes itself out of the case, and swims to a nearby plant. After the skin splits the adult crawls to the surface to rest for a while, so that the wings can dry.

Adult caddisflies do not fly far from the water and they rest by day hidden amongst the plants or on tree trunks which grow close to the pond. As the light fades they take to the wing, and although they occasionally fly into houses, they are more intent on searching for sweet-tasting liquids. The largest caddisfly has a wing span which is just over two inches (5cms). This species lays a batch of 700 or so green eggs on the underside of plants growing in the water. Within nine or ten days the larvae emerge, feed, pupate and repeat the cycle.

A number of beetles spend the early part of their life as larvae in the water. Although the adults may remain in the same pond where they grew and developed, most are capable of flight from one pond to another. This is how new ponds become 'stocked'. This migration often takes place on humid evenings, with many insects taking off from the surface of the water to fly to new ponds.

The great diving beetle lives in many ponds which are surrounded and shaded by water weeds, although patience is needed to catch a glimpse of the insect. Eventually the creature comes to the surface to breathe; here it pushes its abdomen above the water, fractionally raising its wing covers to enable air to be sucked into a pair of small holes called spiracles. Air is also trapped under the wing covers and amongst the numerous hairs on the rear end of the abdomen. The beetle uses this extra supply of air when it is submerged. There are several species but all are generally no more than 1.5inches (37.5mm) across. Both the adult and the larval stages are carnivorous.

Great diving beetle -
voracious carnivore

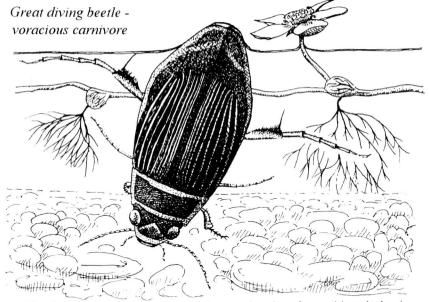

The female has an organ called an ovipositor for making holes in water plants. A single egg is deposited in each hole, and the task is repeated until all the eggs are laid. The adults, some of the most voracious of the smaller pond-dwellers, devour almost any pond creatures within sight, often taking prey much larger than themselves. The larvae, which bear no resemblance to the adults, are also some of the fiercest freshwater feeders. In the water they walk around with their 'tail' ends turned upwards. Ready to change into an adult, the young beetle leaves the water and makes a small shelter in the mud, where it pupates, eventually emerging as an adult great diving beetle. Adults swim by rowing, using the hind legs as paddles.

It is not surprising that another common inhabitant has been given the name 'water boatman' - or backswimmer. Swimming on its back, the insect propels itself along by rowing with the middle pairs of legs. There is a fringe of hairs on the outer edge of these appendages and when moving like this the legs look like paddles attached to a boat. Water boatmen fly freely from one pond to another.

These bugs have sharp beaks, which are strong enough to give a painful nip to humans. Having caught its prey the insect, holds it with its four front legs. Death comes fairly quickly once the beak has pierced it, releasing a mixture of poison and digestive juices. This 'dissolves' the body contents and the water boatman sucks them out. Food consists of small fish, tadpoles and other water insects below the surface, as well as small flies which are pulled from above the water.

The water boatman lays its eggs in spring, the female embedding them in water plants, and numerous miniature water boatman can be seen once the eggs hatch. Within two months they are mature.

Much smaller, and often mistaken for water boatmen, are the corixas. These swim the right way up and are not closely related to the backswimmers. Unlike the backswimmers, they are vegetarians, using hairs on the front legs for sieving water to remove small organic particles and algae, which is then taken in through a short beak.

The water scorpion, another interesting pond-dweller, might be mistaken for a dead leaf at first sight, but once it starts to move its long breathing tube becomes obvious - and this is how it got the name 'scorpion'. People once thought the organ was a sting. The tube, in two halves, is held together by a series of hooked bristles and there are water-repellent hairs at the end to prevent it from becoming saturated. The forelegs, which are modified, are used for holding the creature's prey, which it sucks dry. Scorpions produce their eggs in spring, the female making slits in water plants, where she places a single egg.

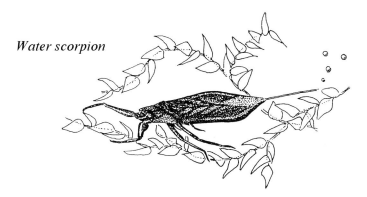

Water scorpion

There is only one true water-dwelling spider in the British Isles, although other species are found by the waterside. Provided there are weeds, water spiders may inhabit the area. Spinnerets produce silk which the spider spins into a web that is different from those made by land-living spiders. The first silken thread is attached to an underwater plant, and the construction of the web continues as more threads are spun. When this is complete, the spider rises to the surface, trapping air bubbles in amongst its body hairs. It then submerges, entering the web from underneath where it releases the air. By repeating the exercise, the web fills with air and becomes bell-shaped, enabling the spider to live submerged for some time. When the supply is exhausted the creature returns to the surface to replenish it.

Whirligig beetles are small insects (5mm across), with black shiny bodies. The creature never seems to get anywhere fast, appearing to spend its life going round in circles. But this is rather a dim view of this versatile beetle, which can dive rapidly and is also capable of flight. Returning to the water, it dives down, using the wings as a crude parachute. It also submerges, taking air bubbles trapped under the wing covers. Its eyes are well adapted for its lifestyle; divided into two, one half watches above the surface, the other below.

The pond skater, or pond strider, can be seen in ponds in summer, when the insects often congregate in large numbers. Its thin body and long legs distribute the weight over a greater area. With a flick of its legs the pond skater propels itself over the surface of the water for a few inches. It does not sink because its light body does not break the surface tension of the water. The pond skater feeds on dead or dying insects. When an insect falls on the surface of the water, sensitive organs on the pond skater's legs enable it to detect the vibrations.

Several species of snail, perfectly adapted to life in fresh water, differ from those living on land. In Britain around 20 of the 36 species of freshwater snails live in ponds. There are two groups: those which breathe through their lungs, and those which take in oxygen through gills. Each gill-breather has a large foot at the rear end of which is a horny plate called the operculum. This is used to close the opening to the shell once the snail retreats inside. Many of these snails can absorb enough oxygen from the water, but some have to travel to the surface to renew their supply.

The other group of snails is made up of the lung-breathers (pulmonates), and these prefer oxygen-rich water. Where the supply is low, they come to the surface to replenish their oxygen supplies. The lung-breathing ramshorn snails are able to survive in water with a low oxygen content because haemoglobin in their blood increases their oxygen-carrying capacity. This is unusual in molluscs, because most have haemocyanin, which is less-efficient at carrying oxygen.

It is intriguing to watch a pond snail gliding effortlessly either over a water plant, or just under the surface film before rising to renew its supply of oxygen. A bubble of air taken into the lung acts as a buoyancy tank, which when released causes the snail to sink to the bottom. The snail's large muscular foot sends ripples backwards as the animal moves forwards. There are two tentacles which point upwards from the head. They differ from land snails in that they cannot be withdrawn. In the great pond snail the tentacles are triangular; in the wandering snail they are long and thin. The shells of fresh water snails are thinner than those of land varieties.

The eggs, which are generally attached to the underside of pond plants, stones or sticks, are enclosed in a gelatinous substance. In some species the eggs hatch in just over a fortnight, and after feeding for a few weeks these snails are ready to mate. Once the young snails hatch, their thin shells make them vulnerable to other creatures living in the pond. If they survive to adulthood they have fewer predators, although herons, ducks, otters and trout may take them. Most pond snails are mature within the year and some, like the common wandering snail, can produce two batches of eggs in a year. Some snail species die in summer, the young snails replacing them.

The great pond snail is the largest of the pond dwelling snails, and may have a 6cm (2.5in) high shell. It is not unusual to see smaller snails, like the ramshorns, attached to the pond snail's shell as the former takes a ride, conserving its supply of energy. Although they eat plant material, they also take other animals living in the pond, and may attack newts, sticklebacks and water beetle larvae.

Of the native ducks which inhabit Norfolk's wide range of waters, the mallard is the best known. During the breeding season the male (drake) and female (duck) sport different plumages. Because of his bright feathers the drake is easier to identify. His head is an irridescent green, enhanced by a white collar, contrasted with a mixture of brown and purple feathers on the breast. The duck is much more drab, being mainly brown but, like the drake, she has a purple wing patch. The male exhibits his breeding plumage from October until June, and his eclipse plumage from July to September, when he resembles the female. Out of the breeding seasons, groups of ducks swim around together probably as a protective measure, making them less liable to attacks from such predators as foxes.

The nesting site is usually on the ground, concealed beneath reeds, or other clumps of vegetation. The nest is constructed of leaves and grass and the female lines it with down from her breast. The first to lay will have eggs in the nest in February, the last by May. A clutch consists of anything from 7-16 eggs, which are incubated for 28 days by the duck. The ducklings leave the nest soon after hatching and their safety is generally the sole concern of the female. It will be some 45 days before they can fly.

Mallard -
Britain's most
common duck

DOWN BY THE RIVER

*Dragonflies, darters and hawkers - The kingfisher, a 'living jewel' -
Waterside plants - Water voles and otters.*

A quiet stroll by a river or stream during the summer will reveal some of the wildlife which occurs there. Dragonflies, water voles, ducks, moorhens, coots, all may be seen provided we have the patience to watch and wait. And in the water, minnows and sticklebacks glide silently by, or fall prey to the bright fast kingfisher or killer heron.

Dragonflies fall into two main groups: the true dragonflies and damselflies. The damselfly - or demoiselle - is often mistaken for a dragonfly, but unlike its powerful flying relatives, it has weaker wings. The bodies are often brightly coloured and include reds, greens and blues. Although the wings may be transparent, they are often tinted with the most beautiful of colours. Damselflies are usually smaller than the dragonflies, having narrow bodies, with four wings, all similarly shaped. A closer examination of these shows that the pattern of veins is different from that of the dragonflies. At rest the wings are folded over the body like those of butterflies.

The true dragonflies can also be divided into two groups: darters and hawkers. The latter are the most powerful, having long, lithe bodies and large wings, some three to four inches (10 cm) wide, and as the name 'hawker' implies they constantly patrol their territories. Hawkers always rest with their two pairs of wings stretched out flat. The smaller darter dragonflies have shorter bodies and spend less time on the wing, preferring to find a suitable sunny perch from which to make regular sorties in search of food. Darters are less solidly built, and are much slower in their movements, Although some hawker dragonflies are often around until October neither they, nor any other members of the order, survive the winter as adults, but new generations emerge from their watery haunts in the following spring.

The swift flash of a brilliantly-coloured dragonfly, on silent shimmering wings, raises the spirits. Pausing on some vegetation its delicate, vividly coloured wings, reveal one of nature's many masterpieces. These dramatic insects are on the wing from about April to September. Some forty different species live in the British Isles emerging at various times throughout the summer months. The eggs of all species are laid in the water, but how, where and when varies from species to species. Some females lay their eggs directly into the water; others wrap theirs carefully in leaves. Some species, like the golden-ringed dragonfly found in western Britain, have ovipositors, which are used for pushing the eggs into mud at the side of running water. In some dragonflies the male continues to grasp the female until she has laid her eggs, but some desert her after mating. The male common darter stays in tandem with the female, and not only does this stop other males from mating with her, but he is also able to give her a 'lift' when she has completed each egg-laying session. Many - perhaps the majority - of eggs never hatch, because water insects, birds and fish, take the eggs for food.

The aquatic larval stages and the adults are completely different. The larva has no bright wings although, like the adult into which it develops, it is a voracious carnivore, often staying still on a stream or lake bottom until it is ready to feed. Having pinpointed its prey, it dashes forward, pushing out a mask at the front of its head and latching on to its prey with previously concealed curved hooks.

The dragonfly - a territorial insect

The mask, with its hinged arrangement, can be folded away under the head when not in use. The larvae of some species stay in the water for up to two years before they emerge to reveal the incredible transformation from dull juvenile to resplendent adult. As with caterpillars, the dragonfly larva sheds its skin at regular intervals.

Fully fed, the larva slowly emerges from the water, and makes its way up the stem of a suitable water plant. Having found a satisfactory site, it rests ready to make the change from larva to adult. In some species the change is almost immediate, taking place in a few minutes; in others it may take some hours.

The county is home to one of the rarest dragonflies in the British Isles. The Norfolk hawker now inhabits a few dykes around the Norfolk Broads. Measuring some 2.5 inches (67mm) in length, it has a wingspan of nearly 4 inches (93mm). Sometimes mistaken for the brown hawker, the Norfolk species has some distinguishing features, which include a triangular orange patch near the base of the hindwings, known as the membranule. There is also a yellow triangle on the abdomen. The species was probably never very common, needing a specialised habitat which confined it to the Fenlands, and now its last haunts are around some of the less disturbed Broads and their associated dykes. After mating, the female lays her eggs on floating vegetation, and the larvae spend two years in the water, the adults emerging to be seen on the wing from June to July, although in exceptional years this period may be extended either side of this.

There are many more animals and plants down by the water, from the microscopic forms to the much larger species. Successful observation requires patience and concentration, but with such a wide variety of life these attributes are well worth cultivating.

The kingfisher is another inhabitant of the waterside, and is often described as a 'living jewel'. Above, it has dazzling, brilliant, green-blue plumage, which gives way to a cross between orange and chestnut underneath. White feathers highlight the neck, and there are similarly coloured patches, one on either side. The kingfisher can never be mistaken for any other bird as it flashes along a stream or river just above the water's surface. But it often remains unnoticed, as it sits statue-like on a tree or post somewhere above or close to the water. Suddenly, and with amazing agility, it dives towards the water, scooping up a small fish from just below the surface, returning to its 'lunch post'. But the bird does not always dive from a stream-side vantage point; sometimes it waits on hovering wings just above the surface of the water, picking up its food from here. Taking the fish from the water does not generally kill it. Instead the bird flies back to its perch, where it strikes the animal several times against the object before swallowing it whole - always head first. If the tail went down first, the scales would open and the bird would choke.

The kingfisher's nest is a burrow in the bank of a stream, river, lake or gravel pit. Both male and female birds fly consistently at the same spot in the bank, gradually loosening the soil with their bills. Having formed a ledge, the birds have a perching place for continuing their work. They excavate a metre-long tunnel, before producing a nesting area at the end, which is not lined, except for regurgitated fish bones. Usually six glossy white eggs are laid any time from April to August. Both male and female birds share the incubatory chores for about three weeks. When the young hatch, feeding is the responsibility of both parents. For between three and four weeks a continuous and increasing supply of fish is brought to the nest site. The nest hole becomes more obvious as time passes, with increasing white lime streaks. It has been estimated that out of every nest only a quarter of the kingfishers live long enough to breed.

Plants, both in and close to the water, add much to the variety of shape and form. One of these, the meadowsweet, is in bloom from June to September. The small flowers occur in large numbers and are borne together on the end of a fairly thick stem, which sub-divides producing a number of irregular-shaped heads. The name is an apt description, as the plant exudes a rich, sweet fragrance, which attracts large numbers of insects. But meadowsweet is a corruption of a former name - mede-sweet - because the flowers were used to flavour ale. During Queen Elizabeth I's reign the plant was used as an 'air freshener', being strewn on floors to keep smells at bay.

Another bright flower of spring and summer is the yellow flag, a member of the iris family, which is evident from the yellow, iris-like flowers. And it is not surprising that poets noticed it too. Gerard

Manley Hopkins described, 'Camps of yellow flag flowers blowing in the wind, which curled over the grey sashes of the long leaves'. The rhizomes yield a black dye and ink of the same colour. In some places it was hung outside buildings to keep evil at bay. And one Fenland doctor suggested that if the seeds were roasted and then ground they made an excellent healthy drink not unlike coffee.

The decline in larger species like the otter is cause for concern. But a small, often neglected, aquatic mammal has also decreased dramatically. A sudden, yet gentle, 'plop' in the water usually heralds the arrival of an animal which is seldom seen. The relatively shy water vole is rat-like in appearance, hence the country name of 'water rat'. It often reaches eight inches (20cm) in length and may weigh as much as ten ounces (300g). In the water the mammal is a superb swimmer and after the initial plop, little is likely to be heard, but it can be followed by watching the slight ripple on the surface of the water. The water vole comes on to dry land to feed, nibbling water plants, as well as the bark of trees growing by the water's edge.

Another mammal which also occurs along the banks of some streams is the water shrew. Not unlike the pygmy and common shrews, it has a special adaptation for its aquatic life-style. A line of hairs on the underside of the tail helps the mammal when swimming, the tail being used as a rudder. The water shrew has a wide range of diet including frogs, tadpoles, small fish, as well as other aquatic creatures. The mammal also comes out on to land, and takes worms, insect larvae and spiders.

Mating takes place from spring to summer, and about twenty-four days later five or six young are born in a nest below the surface of the ground. There are generally two litters a year. Life is short for this mammal and most will be dead within a year.

Once relatively common along Norfolk's rivers, the otter has sadly become a rarity. Fortunately, careful conservation is encouraging the creature back to some of its former haunts This elusive mammal needs secluded, quiet waterways with plenty of bankside vegetation, something which disappeared when rivers were straightened and the adjoining land cleared.

The otter is related to both stoats and weasels and, like these mammals, it has a long, lithe, muscular body, which ends in a tail suitably adapted for use as a rudder. At the other end the head is small, with extremely sensitive whiskers. Superbly adapted for an aquatic lifestyle, its fur acts as both mackintosh and duvet, ensuring that the otter not only stays dry, but also remains warm. Before it dives, it takes a deep breath, and naturalists have heard some otters 'gasp' before they make the plunge. The air is taken into large lungs, enabling the mammal to stay under the water for anything up to four minutes, during which time it can swim about a quarter of a mile.

Solitary creatures, otters are reckoned to need at least a fifteen mile stretch of river for their territory to provide them with enough fishing opportunities. They have a number of daytime resting places which will be used at random.

Careful conservation measures, an improvement in the quality of river water, together with the dedication of a number of naturalists has ensured that the otter, although still rare, is making a welcome return.

LIFE IN COASTAL AREAS

A coast of infinite variety - Conservation bodies - Wildlife on the strandline - Effect of gales - Lugworms - Insects of the shore - Plants and birds of the saltmarshes

One of the most fascinating areas of Norfolk, the coast has innumerable 'special' areas, and in 1968 Norfolk County Council had the foresight to designate 451 square kilometres (280 square miles) between Holme-next-the-Sea and Weybourne an Area of Outstanding Natural Beauty (AONB). This is undoubtedly one of the wildest and most inaccessible areas of coastal marshland anywhere in the British Isles. In 1975 the Council also declared the remarkable coastline between Holme and Kelling a Heritage Coast, which is one of the finest examples of coastline anywhere in Europe. There are high dunes which protect and hide areas of sand flats, shingle beaches and a bewildering maze of saltmarshes and creeks, their silvery fingers stretching as far as the eye can see, glinting in the late evening sunlight. The shape of creek and marsh changes with every incoming and ebbing tide. Here too are ever-shifting dunes. Within this wide variety of habitats there is a unique haven for wildlife and a valuable habitat for a range of plants with a diversity of shape and form unequalled anywhere else in the country.

Other areas have been made into nature reserves, including Blakeney Point (see page 92), the first in Norfolk, and Scolt Head Island, world-renowned for its terneries. The Norfolk Wildlife Trust takes care of other important places, and English Nature, the RSPB and the Norfolk Ornithologists' Union are also responsible for different parts of this unique environment. The 39 mile (69 km) Norfolk Coast Path, which links with the ancient Peddars Way near Hunstanton, runs from Holme-next-the-Sea to Cromer,

The Wash is the largest National Nature Reserve in England, harbouring brent geese, dunlin, shelduck, oystercatchers, redshank and curlew in the saltings. In the inland fields, lapwings gather, often interspersed with smaller numbers of golden plovers and flocks of twite. During the 1992-93 Winter Wader Count the international

60

Above: *Kingfisher*

Photo: Martin Smith

Right: *Meadowsweet*

Photo: Gillian Beckett

importance of the Wash was proven beyond doubt. No fewer than 353,017 waders were recorded in the area, the first estuary to top the one third of a million mark. Beneath the apparently lifeless mud a range of molluscs and worms flourish, providing the waders with rich pickings.

A walk along any Norfolk beach reveals a strandline, which seems to stretch almost endlessly, as far as the eye can see. This line of debris consists of flotsam and jetsam brought in by successive tides A high tide, followed by a succession of low tides, leaves a series of these lines of 'rubbish'. The strandline, newly deposited after a high tide, consists of both plant and animal life. A closer look gives a fascinating insight into the life of the shore and shallow seas around our coast. One of the most common animals in amongst this material is the crab - most commonly the shore crab. Some will be dead, but others are very much alive, revealing something of the intriguing variety of form. This supply of food attracts a range of animals including birds, rats and even other crabs. The softer parts of the dead crabs are removed, leaving the hard exoskeleton. Successive tides drag the discarded material up and down the beach, reducing much of it to unrecognisable fragments.

Shore crab -
food for many
animal visitors

The plants and animals washed up on the strandline give an indication of the passing seasons. Many of the seaweeds, like many of

62

the common inland plants, are annuals. Although their remains are usually washed up in early autumn, severe storms at other times of the year also throw them on to the shore. Like some seaweeds, there are animals which can also be classed as 'annuals', their remains appearing amongst the debris on the strandline once breeding is over.

As with most land animals, spring and summer marks the breeding season for many sea creatures. Large numbers of 'mermaids purses', the egg cases of marine fish - skates, rays and dogfish - end up on the drift line. As the female fish lays her eggs, she wraps long 'tendrils' around the plants, where they remain until the young develop and hatch out. Other egg cases, of the whelk in particular, are also found; these occur in clumps containing many empty cases, and are much smaller than the mermaid's purses. Many are often washed up before the young hatch. Of the large numbers of eggs which are laid, only a few survive and the majority are best termed 'nurse eggs', providing food for those fish fortunate enough to survive. The strandline debris also includes starfish and empty sea urchins, although the latter have generally lost their spines by the time they are deposited on the shore.

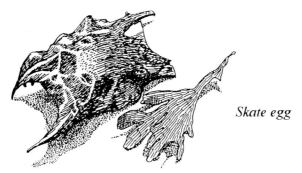

Skate egg

Gale force winds, which are destructive on land, have a similar effect on the sea, ripping up and depositing large amounts of material on to the beach. These high tides may remove many of the burrowing animals, as well as ripping others away from their rocky homes. Seaweeds, torn from their anchorage points, are washed up on the shore. The commonest of these is probably the bladder wrack, a species with small, rounded, hollow lumps at intervals along the length of the fronds, which used to be taken home to forecast the weather.

Above: Otter at Pensthorpe

Photo: Martin Smith

Photo: Martin Smith

Below: Shelduck

There are other species, which often cover rocks lying on the beach. Holdfasts secure them to the rocks under most conditions, but even these tough attachments are no match for fierce North Sea gales. These patches of seaweed remain damp for long periods, providing natural shelters for many seashore creatures even when the tide is out. Severe storms bring in other debris, including wood from the hulls of long-forgotten sunken vessels and timber lost overboard from ships during gales. Having been in the sea for some time the wood is often pitted with small holes made by the ship worm.

Once the tide has receded, many beaches are covered with worm-like heaps of coiled sand. These are produced by the lugworm as it pushes sand upwards during its burrowing activities. The creature lives in a U-shaped burrow and is often sought as bait by fishermen. The common cockle also spends much of its time beneath the sand and empty shells can be found on many beaches. It has a fascinating mode of transport. A long, single foot pokes out from the shell, and the cockle uses this to move from one feeding place to another. Few people see the cockle 'on the hop' because it tends to feed when sea water covers it. The foot is also used as a spade when the creature digs its burrow. To bury itself, the cockle pushes its foot into the sand, and then 'thickens' up the end which helps to pull it out of sight. By being concealed it is less likely to be caught by other sea-dwelling animals like starfish and crabs. Under the sand it feeds by pushing out a thin siphon, and sucks in water containing small amounts of food.

Coastal areas are not always ideal habitats for wildlife, and this is especially true for small winged insects. Off-shore breezes often carry large numbers over the sea where they drown. The area of sand covered by the tide varies from day to day and season to season, and insects landing on the beach may be swept away by the incoming tide. Some species venture into the sea, but few breed there. Many come to feed on the strandline material, and both resident and migratory insects may be seen. The saltmarsh provides a slightly more favourable habitat for some breeding insects, like salad bugs, flies and springtails. Dunes are even better habitats and at Holme Reserve around four hundred species of invertebrates have been recorded.

The saltmarsh on the same reserve yielded more than thirty species, and as many again may be found during migratory activity. At irregular and unpredictable intervals large swarms of butterflies make their way across the North Sea and, like the birds, take the opportunity to rest in coastal habitats. The dunes are often alive with these insects on a warm sunny day, although all are not migrants.

Whereas rockier shores attract a great variety of other small wildlife, Norfolk makes up for this by providing large numbers of particular species. These include estuarine snails and shore hoppers. Wrecks and breakwaters and other large obstacles, may provide one of the few relatively stable habitats for marine life along our coast.

One of the most fascinating - and generally inaccessible - areas is the saltmarsh. Yet in spite of its almost inhospitable nature, it is an important habitat, providing a barrier between the land and the sea, and it is this conflict which presents an ever changing picture. No two incoming or ebbing tides are the same, and the area of saltmarsh covered - or uncovered - varies from day to day and tide to tide. It is because of the unpredictable nature of the marsh that it has remained something of an enigma to many people - but to the advantage of wildlife. So, constantly scoured by successive tides and seasons, the saltmarsh is one part of the coast which has attracted both wildlife and naturalists for a long time.

Plants become adapted to the changes which take place, and the succession which occurs on sand dunes is also characteristic of the evolution of the saltmarsh. The major feature of saltmarshes - and especially those nearest the sea - is the high concentration of salt in the water. This prevents many plants from taking in water, a safety aspect to ensure that plant cells do not burst as the concentration in the cells increases. Many of the saltmarsh plants have succulent leaves which store water, and some of the earlier colonisers include glasswort and sea aster. Not only are the leaves succulent, but the surface has a waxy nature which prevents excessive water loss.

The glasswort is a true pioneer and one of the first plants to appear and establish itself in small patches of bare mud. If seeds land where they are exposed to the air they may begin to germinate. It is the first plant of the saltmarsh habitat as it is formed and once it starts to grow it traps more silt, increasing the height of the land and enabling other plants to come in. To survive in these difficult conditions the leaves have evolved into fleshy scales. Dark green at first, the plant goes through a series of colour changes, starting with yellow, and eventually taking on a pink or reddish tinge. To those who have a taste for it, the plant is the 'samphire' which is still gathered from many marshy areas along the Norfolk coast from June to September and sold fresh on various markets.

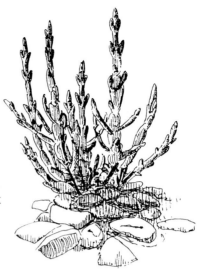

Glasswort (samphire) - a local delicacy

But for those who cannot do without it for the rest of the year, it can be preserved by pickling.

The sea aster is found in the wetter areas of the marsh, often growing along the edges of the creeks. A look at the mauve or yellow flowers soon reveals the plant's close relationship with the garden Michaelmas daisy. Where there is well aerated soil, sea purslane becomes established, and its unmistakable leaves have a silvery-white sheen to them. A cursory glance shows that the stems, which are on the surface, root easily, sending up a number of branches. Where conditions are ideal, the plant forms an impenetrable mass which prevents other species from becoming established. In bloom from July to October, the flowers are very small.

Other plants to gain a roothold include sea plantain and sea arrow grass. But in some areas these plants may be absent, having been ousted by the cordgrass, although this is less invasive than it used to be. Spartina, a hybrid of English cordgrass and an alien, is a rampant coloniser of the marsh, its creeping stems making rapid progress through the mud. Where spartina occurs, the marshes are bereft of other species and colour.

A superficial glance at the saltmarsh may suggest that is an uninhabited and forbidding place, but such an assumption could not be further from the truth. Although insects are generally unable to tolerate the exacting conditions which the saltmarsh offers, it is a haven for a number of birds. The oystercatcher is a frequent visitor, as are redshanks, herring gulls and lapwings. Close to the high tide mark several species of duck may be encountered - depending on the season. These include teal, wigeon, pochard, mallard, shoveler and shelduck. Lapwing find suitable nesting places in amongst the vegetation, and the redshanks, which nest low in the marsh, have natural 'outlets' for their chicks, which float away on an incoming tide.

The lower marsh is inundated twice a day by the sea which not only cleans the area, but also brings in fresh supplies of food. The plants which are established here - Enteromorpha and eelgrass - offer many small sea creatures, such as shrimps, snails and worms, a hiding place. But the presence of these small creatures means food for other creatures, which is eagerly sought out by many waders and ducks, and the plants are also favoured by wildfowl.

Next time you stroll along the coast, take a long, lingering look, you may be surprised at the variety of shape and form which you come across.

Holkham Woods in autumn

Photo: Nicolette Hallett

AUTUMN

AUTUMN WILDLIFE

Autumn food - Bats and hedgehogs - Spiders - Pond-life -
Butterflies and moths - Amphibians and reptiles - Mammals -
Migrant birds

'Behold congenial autumn comes
The Sabbath of the year!'
John Logan - 'Ode to a Visit in the Country in Autumn'

As a predominantly rural county Norfolk provides innumerable animals with valuable supplies of food. As summer gradually gives way to autumn, the grass seeds ripen, providing a rich harvest for birds in the county. Flocks of chaffinches hang over the plants taking their fill, their highly efficient blunt beaks cracking open the seeds. They raid hedgerows, verges and waste ground in their hungry ravages, and at the same time these seed-eating birds offer a useful service to the farmer, removing seeds which would otherwise grow into weeds.

The chain of life links plants and animals because the latter depend on plants for their food. As autumn approaches, the countryside produces its harvest for wild animals. For much of the year there is usually plenty for the wild creatures in the countryside, but autumn provides almost as much food as the animals can eat. Many use this extra supply to build up a layer of fat to help protect them against the forthcoming colder weather, and as a source to draw on if there is a shortage of food. Other animals, like mice, squirrels and jays, collect and hide a store of food, some of which will never be found. Squirrels do not hibernate, but may stay in their dreys for brief periods if the weather is unfavourable, although unless they feed regularly they may die.

Year after year most oaks produce masses of flowers in spring. After the female flowers have been pollinated, gradual changes take place until the familiar acorns appear in the autumn. Towards the end of September and into October the ground beneath the trees, besides being littered with discarded leaves, is often scattered with oak fruits. The number of acorns produced by each tree is unknown , but the figure must run into thousands. Yet in spite of these large numbers few oak trees grow in the following year. Apart from the animals already mentioned, other species are also partial to the oak fruits, including pheasants and wood pigeons. These creatures take their fill during daylight hours, but many other creatures, including rabbits, voles and mice, feed on them at night. Fallow deer add the fruits of the oak to their diet. So when everything is considered it is not surprising that few acorns are left to grow into trees - maybe just some of those which the squirrels or jays hide and forget about!

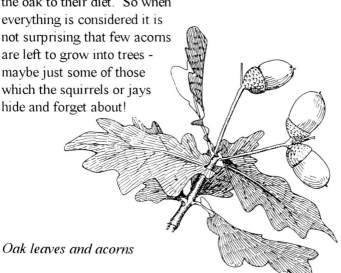

Oak leaves and acorns

Animals living in the British Isles are less fortunate than those in tropical parts of the world. Here our winters may present wildlife with a difficult time, but both plants and animals prepare for this severe season in various ways. Some creatures produce an extra growth of fur, and other animals, like the earthworm, dig deeper into the soil when the top layers become frozen. Yet different methods are adopted by other animals, which 'retire' during these unfavourable months. Where creatures hibernate, the preparations vary from species to species, but all have one basic aim, to lower the metabolic rate to conserve energy. All these preparations begin in late summer and are completed by the autumn.

Although Norfolk's winter weather may be unfavourable, it is not as severe as in some other parts of the world. This is borne out by the fact that only a few mammals hibernate, including the various bat species and hedgehogs. Both build up extra reserves of fat to provide them with a supply of energy while they sleep. Some is stored just under the skin, the rest further inside the animal's body. This extra layer of fat provides the mammal with a supply of food. The outer layer helps insulate it against falling temperatures. Hedgehogs prepare winter sleeping places during the autumn, and studies have shown that there are usually more nests than hedgehogs. This suggests that each builds an additional 'house' in case of emergencies. A dry hedge-bottom or a hole in a tree is a favourite site. The nest is lined with various materials, although leaves usually make up the biggest percentage, presumably because these are waterproof and act as an insulating layer. Ready to succumb to its winter's sleep the mammal enters its nest. Some hedgehogs may come out during December to feed, but this depends to some extent on temperature. Any young hedgehogs which have not reached 500g (about a pound) in weight are unlikely to wake up from their winter sleep, because they have not put down the necessary layer of brown fat.

Hedgehog - useful pest controller

As insectivores bats are without a supply of food during the autumn when insects also disappear from the scene. Bats generally hibernate in the same sort of places as they use for day-time sleeping quarters when they are active. Winter quarters include hollow trees, roof lofts, and outbuildings. The hibernatory habits of bats are not as well known as those of hedgehogs, but mammalogists tell us that extra fat brings about a drowsiness inducing sleep. Some species manage to increase their body weight by over a third. Research has also shown that it is not just a drop in temperature which causes the animal to hibernate. Bats have a remarkable means of overcoming these adverse conditions. The body temperature falls, so that it is similar to the surroundings. This means that the colder it gets, the colder the bat gets, resulting in an ever-decreasing metabolic rate. Temperatures are never constant throughout the winter and bats have periodic bouts of activity, followed by those of deep sleep. If the temperature in the hibernatory quarters rises too much, this will result in increased activity, leading to a decrease in stored reserves. When the temperature rises, bats may have to seek new colder hibernatory sites.

The amount of stored fat varies from species to species and bat to bat, but it is unlikely that many have enough to allow them to wake up more than four or five times. If conditions are favourable when they emerge, and some insects have also been stirred into activity, the bats will take some food to top up their energy levels.

When much of nature is preparing to take things easy, it is perhaps surprising that spiders are at their most active, or maybe it is because we notice them more. An early morning autumn stroll reveals an unending succession of the dew-drenched webs of species like the garden spider. Each strand is studded with a mass of glistening droplets of dew. In early autumn courtship takes place, and this can cause problems, because spiders live solitary lives for most of the year.

The male has to tread warily as he approaches the female. Having built a web, she usually hides, but a silken thread leads from the web to one of her rear legs, and this warns her when anything approaches and lands on the web. The male appears to have a special signal for attracting the female's attention, and if he manages to mate he will leave and seek out another receptive female. He repeats the activity, becoming less active with each encounter. Eventually he either dies or is devoured by a hungry female! After mating, the female garden spider lays several hundred eggs, wraps these in a silken cocoon, and then hides them in fissures in the bark of a tree, or under the cross-bar of a fence or gate. There the cocoon remains, unless attacked by

Garden spider -
master web-builder

predators, until the following
spring, when the spiderlings
emerge. At first they are no
bigger than pinheads, but they
grow and will be ready to mate by
the autumn. Most of the adults probably
die, although some may survive over the winter,
the temperature and food supply dictating what happens.

But not all spiders are such bad parents and some take care of their
eggs. Although wolf spiders never make a web, like all spiders they
are capable of producing silk. The eggs are wrapped in a silken
cocoon, which the female wolf spider attaches to her body. After
some five weeks, the young emerge, and although they are free to
wander as soon as they hatch, when danger threatens they return to
their mother and climb on to her back.

During the autumn some pond creatures make preparations for the
difficult period ahead. In winter the surface of the pond may freeze
over for varying periods. But in spite of this icy cover, life continues
below the surface, and the most popular area of the pond is the bottom
layer in amongst the mud. Here a whole realm of life, unseen and
unheeded by the passing world, continues to flourish. Many insect
larvae live here, the adults having died off. Here they carry on their
struggles so that the species survives until the following spring.
Although some animals remain active, others rest in the mud.

Some fish remain active during the winter, including the
stickleback, roach, rudd, pike, and perch, although they compensate
for the unfavourable conditions by living further below the surface,
and are probably less active than at other times of the year. Other fish

hibernate, including tench, carp and bream; all dig themselves into the mud in the bottom of the pond. Here they remain in a resting state until the temperature increases to stir them into activity.

From late spring many butterflies and moths are on the wing. When autumn approaches these insects have three methods for overcoming the difficult period. The majority spend the winter as pupae, the caterpillars seeking suitable places to hibernate. Yet other species like the large white, spend the winter as chrysalides, whereas some, including the small tortoiseshell and brimstone, pass this difficult period as adults. An early warm spell in spring may entice the hibernating adults out for brief appearances, before they return to their winter quarters. If there is a sudden drop in temperature these insects may die before they make it back to the hibernatory state.

Pale tussock caterpillars feed from June to October, and are common in woodland, taking their nourishment from a variety of tree leaves. The larvae are especially attractive, with hairy bodies consisting of creamy-green tufts with black banding which is revealed when the caterpillars stretch out. If disturbed, they tend to curl up into a ball, although the prominent orange spike at the tail end is still visible. Ready to pupate, the larvae spin strong double-shelled brown cocoons which afford good protection during the colder months The moths remain here in a state of suspended animation until next spring.

Amphibians and reptiles are cold-blooded animals which means that the temperature of the body is almost the same as that of the surroundings. As the atmospheric temperature drops, so does that of the amphibian or reptile's body. When this temperature reaches a certain level, the animal automatically hibernates. Frogs bury themselves below the surface of the mud, where the body activity, though barely discernible, continues. Toads prefer to hibernate in drier situations. Lizards find cracks and stones where they pass the winter. Most newts hibernate on land in winter, as do snakes, which seek sheltered spots under fallen branches or in hollow tree trunks.

Throughout spring and summer the countryside has been clothed by green leaves from the numerous trees which grow in the county. In autumn it is the rich golds, reds, oranges and browns which make this season of the year so beautiful. These brilliant, vibrant colours are

Left: *Gannet
in flight*

Photo: Roger Tidman

Below: *Common seals
at Blakeney*

Photo: Martin Smith

present all year round, but are masked by the green colouring - chlorophyll. These changing colours, coupled with the falling leaves, remind us that autumn has arrived. Not all trees lose their leaves at the same time. Initially a few start to fall, drifting idly to the ground, but gradually the pace increases until winds and frost help to send them cascading down in ever-increasing volumes. Some trees become bare; others may still be almost fully clothed. The leaves have served their purpose of making food and, once dead, they are discarded. Young oaks often hang on tenaciously to their shriveled brown leaves right through the winter, until the new growth forces them off the tree and on to the ground. For a few months the tree is safe as it rests with all the parts above the ground sealed against adverse weather.

Although many plants and animals have prepared for autumn, there are large numbers of other species which do not. Mammals, like the fox and rabbit, manage to continue an almost normal existence. Like many other animals, the rabbit is 'aware' that there is likely to be a shortage of food. With plenty in the autumn countryside, these animals eat much more, building up a store of fat beneath the skin, to tide them over when it is impossible to find fresh food.

Some birds, unable to survive the climate in this county during the winter, set off in late summer or early autumn on their migratory flights to their winter quarters, making the long journey to warmer, more hospitable, lands such as Africa. Most insects also disappear from the countryside in autumn and most birds which rely on this food set off to spend the winter in Africa, where these invertebrates are still active. These include the swifts, swallows, martins and flycatchers.

Although the British climate is unfavourable for some birds, it is much kinder to others than that in their own lands, and so, as some species depart, others come to our shores for the winter. Birds are able to adapt to cold weather, except in very severe conditions. By puffing up their feathers they trap warm air which helps insulate the body. Smaller species, like wrens, are more vulnerable and are likely to perish more quickly in very cold conditions. In continuous periods of severe frost, as experienced in the winter of 1962-63, large numbers of birds die because they are unable to obtain food and water. Birds which rely on the soil for their supply of food find feeding difficult - if not impossible - when the ground is frozen.

If the coast is an attractive place in the summer, it is just as much a magnet in the autumn when large numbers of migrant birds arrive. One of the most spectacular of these is the gannet. With a 6ft (1.8m) wingspan, it can hardly be mistaken for any other species. On the wing, its long neck and pointed tail are distinguishing features. The wings, although extremely powerful, are quite narrow for such a large bird. Young gannets take five years to acquire the adult plumage, and the black and white blotched feathers of the young birds are gradually replaced by white plumage and black wing tips. Most food is caught by diving into the water and the bird can often be seen plummeting into the sea from heights greater than 100ft (33m). The nostrils have evolved to prevent the entry of sea water. When the bird dives towards the water from a height of 300 feet (100 metres), it hits the surface at around 60mph (97kph).

Although the turnstone can be seen feeding during the summer, it is more common in the autumn. Appropriately named, the bird spends a great deal of its time turning over stones in search of food. The eyes are well positioned on the head, so that the bird can see behind as well as in front, advantageous when you spend much of your life with your head close to the ground. Regaled in its breeding plumage the bird has rich chestnut feathers, with a darker breast and black to the upper part of its body. After breeding, this coat moults and is replaced by a much more drab plumage.

The Norfolk coast attracts many other migrant waders, including knot, dunlin, sanderling and green sandpiper. The dunlin seeks out the marshes for feeding, with the sanderling taking to the shore line. The number of knot which spend the winter along the coast fluctuates wildly from year to year.

Just as spring was the time of the year when plants and animals reappeared for the favourable conditions ahead - to mate and produce offspring - autumn is the season when animals and plants prepare for the unfavourable months ahead.

SEALS IN THE WASH

Seals in Norfolk - The breeding cycle - Hazards for seal pups -
Aquatic skills - Population and distribution

The Norfolk Coast, with its many off-shore sandbanks, provides an ideal habitat for both common and grey seals. For many people their first sighting of a seal is when they take a trip off-shore and spot a bobbing head in the water. For others it is an encounter with the mammals as they bask on a sandbank. At low tide Blakeney Point Nature Reserve is one of the best places for spotting these fascinating mammals. With a pair of binoculars trained on the off-shore sandbanks, large numbers of common seals and lesser numbers of grey seals can be seen resting on the golden sand. Each year some young seals are either washed ashore, or follow pleasure boats and fishing boats back to land.

Although aquatic by nature, the common seal often visits sandbanks and, less frequently, the coast, but it has to come to land to give birth to its young. Common seals produce their first young in June with others arriving in July and August. Grey seals are later, giving birth from October to December.

Both grey and common seals have an interesting breeding cycle. Mating takes place soon after the young are born, but because the females have been drained by giving birth and are still suckling their young, the embryo does not develop for some time. Eventually the egg implants itself in the uterus, a form of development known as delayed implantation.

Adult common seals moult in August and at this time the older ones are particularly inactive.   Seals have already paired off and the first courtship activities took place earlier in the year - in April or May - but this stopped when the young were born, and courtship is 'shelved' when the seals are moulting, resuming once they have their new coats. There is a difference of opinion about seal-bonding. Some researchers suggest that the males (bulls) only have one mate (cow) each year, remaining faithful to her. Other observers differ, suggesting that the bull common seal has a number of 'wives'.

According to the fishermen in the Wash the greatest number of seal pups are born at low tide on Midsummer Night - the night of 21-22 June. At birth the common seal pup averages around 39 inches (under a metre) in length and weighs about 20 lbs (9kg), although some are larger, tipping the scales at 18kg (40lbs). In the womb the developing pup has white fur, but this generally moults before birth, the under brown fur becoming prominent. Most common seals come to the sandbanks to have their young, although some are thought to give birth in the water.

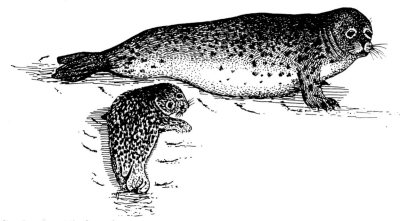

Seal pup with female

For the first few hours the young seals lie helpless on the sandbanks making few attempts to move. As with all mammals, mother provides them with a supply of milk which they take from her on land and in the water. The female often leaves her offspring and returns to the water if disturbed, although some observers have seen the female trying to push the young seals into the water to avoid any danger. Ungainly on land, it is no wonder that the young weak pups have difficulty in moving. In the water it is a different story, but for the first few days they stay close to the female, feeding frequently. It is at this early stage that the young pups may be distracted by humans who try to make a fuss of them, drawing them away from their parents. It was because of the problem caused by these seal pups that the RSPCA originally set up seal nurseries at Heacham on the Norfolk coast and further inland at Docking. Now seal recuperation is concentrated at the East Winch Rehabilitation Centre. In the early days of the seal rescue operation the RSPCA fed and cared for the pups, returning them to their sandbank homes in the Wash when they were strong enough. Sea-Life Centre at Hunstanton also has a special unit for dealing with stranded and orphaned seals. On the other side of the Wash, Natureland at Skegness has been rearing young seals for many years.

As the pup continues to grow it still has trouble moving on land; it shuffles along using its fore-flippers as levers, and pushing its head into the air, exposing the underparts. After a short distance the pup flops inelegantly on to its belly and, using the hind flippers like oars, makes the next lurch forward. If the seal pup is ungainly and almost comical on land, the water transforms it. Equipped for an aquatic lifestyle the seal swims and dives with grace and ease. Although much more at ease in the water the seal pup faces some danger. When separated from the female a heavy swell may carry it towards land, where it might either become stranded or, more seriously, be dashed against a cliff. Strong currents in the Wash, which have defeated many human swimmers, may also pose problems for the young and, at this stage, relatively weak-limbed seal pup. Some pups which are parted from their parents may never be reunited and so die from hunger or exhaustion or a combination of the two. Luckier pups are picked up and taken to seal rescue centres.

Of the thousand or so common seals which live in the Wash, probably only a few pups die in this way each year. Once separated from its mother the seal pup utters a prolonged, plaintive wail and, except in exceptional circumstances, the female will be able to trace and recover her offspring. Should a pup become exhausted by staying in the water too long, the female protects it by cradling it in her flippers. Exercise is important for the growing pup, and mother encourages her offspring to play with her. The time spent in the water increases, and the continuing exercise develops the pup's muscles. The bull appears to take no part either in the training or upbringing of his offspring. The cow continues to suckle her pup for some five weeks. During the first two weeks the female is very attentive, and this is a particularly playful period. But as the pup grows, the female's interest wanes as they prepare to go their separate ways. 'Games' now consist of finding floating objects to play with. Seaweed, driftwood and almost anything it finds, either in the sea or on the sandbanks, arouses the pup's curiosity. Later it turns its attention to other pups and when it and the other youngsters are exhausted, the young seals rest on the sandbanks.

Nowadays the pups are safe on the sandbanks, but during the 1960's they were in danger from seal-hunters out to kill them. The young seals were easy targets because of their lack of speed on land The enemy struck when the young pups were four months old, and when the coat was said to be at its best. Although many of the hunters were skilful, the less competent caused the seals great distress. Many which were washed ashore around the Lincolnshire and Norfolk coastlines, had been shot. With the introduction of The Conservation of Seals Act in 1970 a close season came into force from the beginning of June to the end of August, and licenses were needed from the Home Office to kill seals. However, the anti-seal hunting lobby was effective and the last licence was granted some 20 years ago.

Unlike common seal pups, the first week or so of the grey seal's life is spent virtually in solitary confinement. During the summer, bull seals spend most of their time in the water. On land all seals are clumsy; in the water they are superb. The fore-flippers, positioned

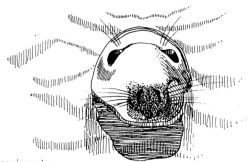

Grey seal swimming

close to the side of the body, streamline the shape. The hind-fins are likewise 'pinned' to the body, with the webbed feet spread. The seal uses these as rudders which guide it through the water. At slower speeds, stabilisers are necessary, and the fore-flippers provide these, being moved in short circular swings to keep the seal on an even keel. The hind-flippers are also used; they open and close in wider circles, and the combined efforts propel the seal slowly forward. In water, seal movement is fish-like, the tail being moved quickly from side to side.

Because of their elusive, aquatic nature, the number of common seals around the coasts of the British Isles is difficult to establish, but estimates suggest that the greatest concentration is in the Wash, the population numbering more than two thousand. Herds of between three and four hundred individuals have been counted, hauled out on the vast areas of banks exposed at low water. A colony of more than five hundred common seals lives in Blakeney Harbour.

Although seal-culling in the sixties destroyed around five thousand mammals, since the 1970 Act numbers have increased again. Although the population of grey seals is much smaller, their numbers have also increased, with some mammals moving from Scroby Sands off Great Yarmouth. These sandbanks are unstable, disappearing and reappearing. This poses problems for the grey seals, because the young must be land-based for the first few days of life. With more stability in the Wash, the mammals have moved around the coast. The first grey seal pup was recorded on Scroby Sands in 1958.

To understand seal movement tags are placed on the mammal's flippers, and by checking these it is known that they travel down from the Farne Islands off the north-east coast to join the resident Norfolk population. When the first grey seals arrived off the Norfolk coast breeding was limited, but pups are now born each year.

Feeding is an important activity, and when the sandbanks are covered by the rising tide, the seals use this period to catch food. At high tide the seals can be seen swimming, their heads just out of the water. Food consists mainly of fish, although molluscs are also taken. Fishing in the murky waters is not a problem because the vibrissae - stiff whiskers - pick up underwater vibrations, enabling seals to locate their food.

Quite naturally fishermen condemn the seals because of the large numbers of fish which they eat. Stomach contents of some seals show that both fish and molluscs make up the food supply in equal quantities. With thousands of seals in the Wash the amount of food consumed is considerable. An old bull grey seal is estimated to weigh more than 450 kg (average about 250kg); common seals tip the scales at 250kg (average 150kg). Females are much smaller.

The waters around the Norfolk coast appear to harbour few enemies, and although the killer whale, a member of the dolphin family, visits North Sea coasts from time to time, and may take some seals, its activities are of little significance to the seal population. Although a great deal is known about the lives and habits of these fascinating creatures of the sea, much more still remains to be discovered.

CREATURES OF THE NIGHT

*Sounds of the night - Hedgehogs, hares and badgers - Barn owls -
Night-flying moths - Foxes at night*

As the day draws to a close, humans prepare for rest, and even though we may take a stroll on a moonlit night, we seldom stay up throughout the hours of darkness. Like many animals of the countryside we are diurnal, working by day and sleeping by night. But as one 'army' of life prepares to rest, another gets ready for work; these are the nocturnal creatures. When we are out on a dark night, it is very difficult to get about, because we are unable to see properly. Unlike nocturnal creatures we do not have eyes which are equipped for seeing in the almost non-existent light of night. For some creatures the hours of darkness seem ideal. It offers shy, retiring animals a sense of security as they come out of their sleeping quarters into a world either clothed in darkness or relatively dimly lit by the moon. Such a landscape ought to provide them with ideal predator-free conditions.But although these creatures are abroad at this time, they are not necessarily safe, because other animals emerge at night to hunt for food - live food in the form of small creatures which abound in the hours of apparent tranquillity and darkness. If we assume that the night is quiet and peaceful, a nocturnal saunter reveals a multitude of sounds. A screech from a high tree as an owl lands warns its fellows of impending danger. Snuffling in the long grass might herald the activities of a hedgehog. Grunting from the undergrowth could mark the emergence of a badger from its sett. And then there are the haunting sounds which echo and re-echo through the wood and across the countryside as dog fox barks at vixen.

The night-time temperature controls - at least to some extent - the activity which takes place. If the autumn air is mild, then moths flit silently on the wing in their search for food. Overhead a bat glides on its quiet path in search of food in early autumn to top up its fat reserves before hibernation. The early autumn evening atmosphere may still be tinged with the scent of the odd flower which exudes its fragrance upon the still, night air. The seasons of the year mean changes too; a frost on the leaves gives some plants a white, almost ghost-like appearance. Or dew may cover the grass, which glistens like twinkling stars in the bright moonlight. If there is no wind, then silent, shimmering mists, white and clammy, may collect in the hollows of undulating countryside. And while one world is still, another continues, unheeded, unwatched and unimpeded by humans.

The nocturnal hedgehog has several local names, including 'earthpig' and 'urchin'. During the day the animal sleeps, usually on a bed of dry leaves in a hedge bottom. As dusk falls it leaves the relative security of its daytime haunts to feed. If undisturbed, the hedgehog hurries along, but at the slightest sound it freezes on the spot, drawing the spines to the top of its head. Should real danger threaten, it rolls itself into a ball, withdrawing the head and all four legs. The spine-covered coat curls around to protect the mammal's more vulnerable parts. Although hedgehogs are thought to have poor eyesight they are compensated with acute senses of hearing and smell.

Once on its rounds it is not a particularly fussy feeder, and takes a wide variety of food. For preference, slugs, insects, snails and worms form the major part of its diet, but it also eats berries and occasionally eggs. Like other mammals which hibernate, it will eat more than it needs and so puts on reserves of fat to see it through the winter.

The hare, another creature of the night, is abroad from dusk to dawn, although it is also often seen during the day, bounding across fields. Unlike the rabbit, the hare never makes a burrow, but sleeps concealed among plant material. Here it makes a depression which is the exact shape of its body, and possibly because of this, it is called a 'form'. These daytime resting quarters are generally no more than 4 inches (10cms) deep. Where forms are in long vegetation, only the creature's head and ears protrude. These forms can sometimes be found because the hare uses the same route from its form when it goes off to feed. The hare also has keen senses of smell and hearing. Its whiskers are constantly on the move to detect any scents which might suggest that danger is near. If the hare suspects its safety is in jeopardy, it makes additional surveys by standing up in the form, so that it can see around it. When threatened, the adults emits a shrill screech. The hare feeds at night and is a true vegetarian.

Few people have observed the badger in Norfolk, for two very good reasons. First, it is quite rare and second, it is nocturnal by nature. In many places it was known as 'brock' - a word which aptly means 'parti-coloured', and places like Brockdish near Diss probably get their names from badgers. The flat nature of much of Norfolk's landscape is one of the reasons why there are few badgers, coupled with the fact that they were persecuted in the past. Badgers do not hibernate and emerge regularly from their setts, although the time varies from one sett to another. During the autumn and winter it is usually an hour after sunset, and in spring and summer brock emerges earlier. If the badger suspects danger before or immediately after it has left the sett, it may return and stay below ground for the night.

The badger belongs to the group of mammals known as Carnivora - the flesh eaters. The teeth are well adapted for catching and killing quite large animals, so it comes as something of a surprise to discover that its diet consists mainly of worms and other soft-bodied food. From time to time badgers take young rabbits, chickens and even lambs. Small mammals, including moles, mice and voles, as well as frogs, may also form part of its diet. Brock will also raid wasps' nests to feed on the larvae. Some plant material is also taken, including bulbs, grass, blackberries, acorns and fallen apples.

Badgers

A number of legends have become associated with birds found abroad during the hours of darkness, and the barn owl is no exception. Its silent flight often frightened superstitious country people and because of the light-coloured plumage it appeared ghostly in the dim moonlight. Changes in agriculture and countryside practices, together with continuing urbanisation, increases in the road network and loss of feeding areas are all thought to contribute to the owl's decline. Sites favoured for nesting - like hollow trees and old buildings - have all but disappeared. From 1982-85 the Hawk and Owl Trust surveys of barn owls showed that the population had fallen by 70% since the 1930's. As a result of this the British Trust for Ornithology (BTO) and the Hawk and Owl Trust (HOT) are joining forces to carry out a new survey, the pilot study having taken place in 1994.

With the onset of twilight the barn owl leaves its daytime refuge, flying low over the Norfolk countryside. The owl's flight path remains fairly constant, and it can be seen flying over the same area night after night, having established its territorial rights. Owls circle over a region, suddenly plummeting to earth on silent wings to grasp a victim in sharp and powerful talons. With its prey safely collected the barn owl returns either to its regular roosting site or to its nest. Owl roosts and nests are conspicuous because of large numbers of pellets which are deposited in the area. The owl swallows its prey whole, but bones and fur are indigestible. This material is wrapped up in the fur enabling the owl to regurgitate it more easily. By collecting and unravelling these pellets a picture of the owl's food supply can be built up. The owl is a friend in the countryside, taking small mammals which, if they increased out of hand, would cause problems for farmers and landowners.

The owl nests in a hole, the eggs generally being laid on an accumulation of pellets. Sometimes she may tear up the pellets, or even bring in nesting material. The job of incubating the eggs is the responsibility of the female, but the male keeps her supplied with food. For around five weeks she sits on the eggs. After hatching, the young remain under the careful eye of both parents for between two and three months. Then they are ready to leave the nest to find their own territory. Once established, it is theirs for life, and here they stay to feed and mate.

During the warmer months of the year, moths are on the wing at night if conditions are favourable, but as autumn tightens its grip on the countryside, fewer species are found. Angle shades moths are still on the wing, but they are not particularly conspicuous because their green and chestnut markings effectively blend in well with the autumnal leaves. The merveille du jour is also about at this time, effectively camouflaged when it settles on lichen-covered tree trunks. At least one species of tussock moth - the vapourer - may be out and about during the early autumn. All tussocks are very hairy, a feature which also applies to both larvae and pupae. The eggs are laid in the autumn, the moths emerging from the black pupae between July and September. The male is often seen in flight during the day.

Numerous other species occur in Norfolk, but some of the most interesting belong to the group known as hawk moths. These large insects have thick, plump bodies and powerful wings, which enable them to attain relatively high speeds. The strong flight is an indicator that hawk moths, like many birds and some butterflies, are migratory by nature. Like the adults, the hawk moth caterpillars are large and very conspicuous. One of the most noticeable is the privet hawk moth, its bright green segmented body surmounted by a hook-like appendage on the posterior end. The adults are on the wing in May and June, coming out as soon as dusk settles over the Norfolk countryside, searching for nectar which is produced by many flowers.

When a resting hawk moth is disturbed it is capable of almost immediate flight. It beats its wings vigorously for a few seconds before taking off. This warms up the body, because the resting temperature is lower than that needed for flight. This is one of nature's strategies to ensure that the moth conserves as much energy as possible for when it needs it most.

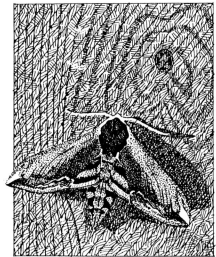

Privet hawk moth

Females lay their eggs on privet and lilac, and the emerging caterpillars feed from July to September, then make for the soil where they bury themselves several inches below the surface. The pupae are either chocolate brown or chestnut and remain here until next spring, although many probably fall prey to tunnelling moles.

*The much
maligned fox*

The fox is one of the largest creatures to be abroad at night. For centuries foxes have been persecuted by country-dwellers, and especially gamekeepers. Others have been killed because of their attacks on poultry. Although mainly nocturnal, the fox is sometimes seen during the day, especially in wooded areas where it is unlikely to be disturbed. The female (vixen) and male (dog) live independent lives, only coming together for the breeding season. At rest the fox spends much of its life in its underground home called an 'earth'. Some of these are specially excavated, but others are taken over from rabbits or badgers. The most common call of both dog fox and the vixen is the familiar barking, especially during the winter months, but when the breeding season is imminent the bark is replaced by a raucous scream, which often echoes uncannily in the still air of a darkened countryside.

The main food is small mammals, with poultry and lamb-stealing relatively rare. Earthworms are also a favourite food, and a survey by the Ministry of Agriculture, Fisheries and Food into the stomach contents of foxes, showed that the remains consisted mainly of bank voles, rats and mice, and this was supplemented by a diverse range of food including hedgehogs, snails, frogs and partridges as well as a great deal of vegetable matter.

And so, imperceptibly, dawn slowly creeps over the Norfolk countryside as the sun starts to rise, sending the nocturnal creatures to their rest, ready to return at dusk for another vital nightly patrol.

THE COUNTRYSIDE AT NIGHT

The sun is set, and everywhere is still,
The moon comes up from way beyond the hill,
Soon creatures of the night will start to roam,
A fox slinks out in the oncoming gloom,
And hark! an owl hoots low by yonder bridge,
A bat glides by o'er that far distant ridge

Nocturnal cries re-echo through the wood,
A carcass lies where once a weasel stood,
The badger leaves its sett to hunt its prey,
Returning at the start of a new day.

Death strikes so often in these darkened hours,
A mouse is held in strong and powerful claws,
The air is filled with awesome, deathly cries
Unheeded under dark and threatening skies.

The moon begins to fade, the sun is rising fast,
Nocturnal creatures know that night is past,
And they must to their homes before its light,
Dusk is their day, and day their quiet night.

BLAKENEY POINT NATURE RESERVE

Norfolk's first nature reserve - Formation of dunes and ridges -
Plant growth - Migrant breeding birds - Other visitors

Violent winds and heavy seas, peace and calm with hardly a ripple on the sea - these are the contrasting moods which characterise Blakeney Point. Away from the main road which snakes its way along the Norfolk coast, this nature reserve is undoubtedly one of the most unspoilt and isolated pieces of coastline found anywhere in Norfolk.

Here, undisturbed by the constant flow of traffic along the A149, is an area of great fascination to the growing number of people who are becoming increasingly interested in the wildlife and conservation of our island.

London University has used the Point for scientific study for many years, and it was due to the efforts of the late Dr Oliver, a member of staff from University College London, that Blakeney Point became a nature reserve for generations of nature lovers to enjoy.

Blakeney Point was the first nature reserve to be established in Norfolk, and was given to the National Trust in 1912. The Point and nearby Scolt Head Island were run jointly by the National Trust and the Norfolk Wildlife Trust, but Scolt Head Island was leased to English Nature.

Blakeney Point, which is still administered by the National Trust, covers some 1200 acres (486 hectares) and extends about four miles (6.5km) from the mainland into the North Sea. But it is necessary to take a look at the coastline from Hunstanton to beyond Blakeney Point to see the nature reserve in context. In simple terms Thornham has relatively small sand dunes with lesser dune ridges. Scolt Head has embryo dunes whereas at Stiffkey there are vast expanses of saltmarshes. Seaward of these are to be found ridges which have been built up by the action of the waves. At some places, like Holkham and Titchwell, the deposition has been so great that land reclamation has taken place.

The formation of sand dune and shingle ridge is an extremely

Right: *Barn owl*

Photo: Martin Smith

Below: *Grey seals*

Photo: Martin Smith

complex process, but as in many coastal areas two opposite activities can be seen along the Norfolk coast. These are erosion and accretion. Erosion is the washing away of material from coastal areas, something which is taking place further along the coast at Hunstanton and in the Cromer area. Accretion is the opposite process in which material is built up, which is how Blakeney Point came into existence. This process still continues to 'add land' to the area. Within this environment there are a number of specialised habitats for both plants and animals. Broadly speaking these can be divided into shingle ridge, sand dune and saltmarsh. The whole of the Point centres around the shingle spit.

At first sight, areas of shingle lashed by winds and waves might appear unlikely places for plants to grow. The effect of both the wind and the waves on the exposed area is considerable and it was these forces which formed Blakeney Point. Basically, the shingle spit and shingle ridges are made up of pebbles of varying sizes, some of which may have travelled considerable distances. Marked pebbles dropped in the sea off Cromer have been found off Blakeney Point. Many of the shingle ridges are constantly sprayed with sea water when the tide is in, and this tidal action changes the shapes of the ridges. There are also heavy storms along this part of the coastline and these are responsible for moving vast quantities of shingle from one place to another.

A look at the shingle gives a clue as to the ways in which plants survive. If pebbles from the top layer are lifted up, even on a very hot day, the underlying material is quite wet. The first reaction is that this is salt water, which would be unsuitable for many plants. But tests show that this is fresh water, which has been deposited during changes in daytime temperatures. Although this water is important for plants to grow, they also need food. In the first instance this comes from plant and animal matter carried in on the tides. Over a period of time this material decays, releasing minerals which dissolve in the water to be absorbed by the plant's roots.

The shingle attracts birds which feed and nest on the ridges and they provide additional mineral matter. The areas of shingle which birds visit often have the most luxuriant plant growth, and a surprising feature of the habitat is that a wide variety of plants becomes established. Because of differences in the make-up of various parts of the ridges the plant life also varies. Shrubby seablite is one perennial plant which is confined to the shingle ridges. As its name suggests it is a shrub-like plant with fleshy leaves, and grows more vigorously when covered with shingle. It reaches its northern limits along the North Norfolk coast.

Another perennial, the sea campion, covers many of the ridges, producing a carpet of white blooms during the early summer. The yellow horned poppy also becomes established in such areas, its bright yellow flowers adding a splash of colour to the shingle, with some plants still bearing their flowers into October. These plants are the early colonisers on newly-formed shingle ridges. When the shingle is more mature, other plants start to grow, and the biting stonecrop, a common species in almost any habitat, also occurs here. This plant is often abundant, together with others like plantains, docks and groundsel.

Shrubby seablite

The sand dunes are also of great interest. The sand which makes up the dunes comes mainly from two sources. Much is blown in from the numerous off-shore sandbanks - also known as shoals - which dry out at low tide. The other source is the wide expanse of open beach which is laid bare at low tide. Dunes are formed when sand particles, carried by the wind, are trapped against any object on the beach. Dune-building is speeded up when plants germinate and stabilise the blown sand.

The formation of the dunes, and their size and shape, depends on a number of physical activities, with the direction, duration and strength of the wind being major contributory factors. The amount of sand and the shape of the beach are also important.

Plants, like prickly saltwort, glasswort and sea rocket, growing along the drift line, trap blown sand. They are often responsible for the first embryonic dunes. These early plants have fleshy leaves which can store water, ensuring their survival in adverse conditions. As the accumulation of sand increases these plants can no longer tolerate the new conditions and die off, while other plants take over. Three species of grass - marram, lyme and sand couch - together with sea sandwort, are responsible for the greatest dune formation.

Like the shingle ridges, these first dunes lack water and nutrients, and only a few plants can survive in the difficult conditions. As the dunes become established, both sand couch grass and lyme grass help to stabilise the sand, but it is the arrival of the marram grass which is responsible for the rapid growth of any dune system. All three grass species grow more vigorously when they are covered by layers of sand. The roots of marram grass may go down into the dune system to a depth of some 21 feet (7m). Marram grass spreads by this underground rooting system which, as well as producing new plants, also stabilises the sand.

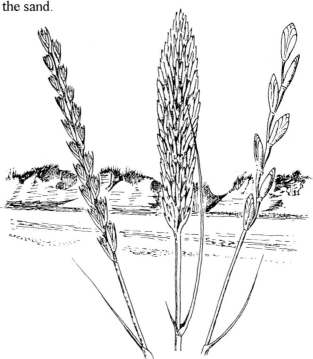

Marram, lyme and sand couch grasses

Above: *Yellow horned poppy growing in shingle*

Photo: Gillian Beckett

Photo: Bryan Sage

Below: *Little tern nesting*

These young dunes, known as yellow or white dunes, give the area its distinctive colour. With little food and water few other plants can survive, but those which do gain a foothold include dune fescue, sea bindweed, sea holly and sand sedge.

As the dunes mature, conditions improve and more plants come in. And, as with the shingle ridges, plants found in other habitats become established. Common garden 'weeds' are recorded on older dunes where the soil is fertile and these include docks, plantains, dandelions and sow thistles.

With its prominent position jutting out into the North Sea, Blakeney Point attracts large numbers of nesting birds which arrive in spring and leave again in the autumn. These sea birds make very simple nests - either a dip or hollow in the sand or shingle is all that is needed. When the young birds hatch, most begin to run around almost as soon as they leave the egg, making an elaborate nest unnecessary.

A colony of common terns takes over part of the shingle spit each year, the numbers present varying from one year to another. In a good season there may be as many as a thousand breeding pairs of terns nesting in areas known as terneries. Their excrement adds to the mineral content of the soil, which in turn helps to increase plant life.

In contrast to the activities of the common terns, the little or lesser terns prefer to nest solitarily. Other species of terns which breed on the Point include Sandwich and Arctic terns, although their breeding activities are less predictable. The Blakeney Point terneries are of national importance.

The oystercatcher nests in amongst the terns. These distinctive birds can be recognised not only from their unmistakable calls, but also from their contrasting black and white plumage. Their 'smart' attire is completed by a distinctive orange beak and pink legs.

Various gulls breed on the Point, the most regular being the black-headed gull. During the breeding season, the bird is characterised by a chocolate brown head.

Other breeding birds include the shelduck. About the size of a large goose, it nests in rabbit holes. The ringed plover, characterised by a broad band of black which runs across the white chest, prefers to nest away from the terneries, often selecting sites where there is some grass cover providing both the sitting birds and the eggs with some protection.

In the winter visiting birds arrive including migratory waders which may either pass through, or may take up semi-permanent residence before they leave for their breeding grounds the following spring.

There are many other species associated with this part of the Norfolk coast. Some are shoreline visitors, stopping off to 'take a bite'; others nest and yet more are residents, probably only being driven away temporarily in severe weather conditions. In the winter, this bleak coastline provides a feeding ground for thousands of waders. And this world-renowned nature reserve, holds the distinction of having the highest number of bird species recorded for any place in the British Isles. Many interesting species are recorded annually; others turn up but once in a lifetime.

Apart from the bird life, there are also large colonies of common seals (see p.79) to be found on the off-shore shoals. Grey seals are also resident off the Point, having arrived from other areas where changing environmental conditions have deterred their breeding activities. The appearance of grey and common seals together off Blakeney Point is unusual because the two species seldom bask together.

Blakeney Point attracts all kinds of people, from the serious botanist and ornithologist, who will find plenty to study, to the casual visitor who can enjoy the unique environment which will stimulate the mind and lift the heart. It is well worth making the effort to visits this delightful part of the Norfolk coast in its autumn glory.

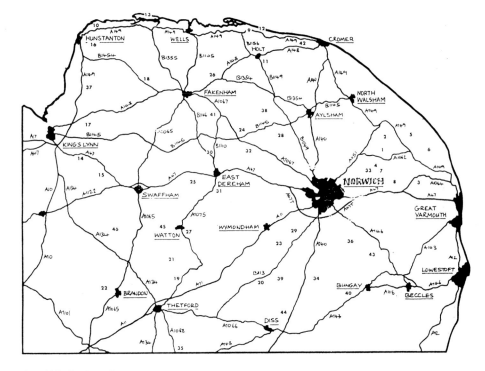

1. Alderfen Broad
2. Barton Broad - National Nature Reserve
3. Burgh Common
4. Cockshoot Broad
5. Hickling Broad
6. Martham Broad
7. Ranworth Broad
8. Upton Fen
9. Cley Marshes
10. Holme Dunes - National Nature Reserve
11. Holt Lowes
12. Salthouse Marhses
13. Scolt Head
14. East Winch Common
15. Narborough Railway Line
16. Ringstead Downs
17. Roydon Common - National Nature Reserve
18. Syderstone Common
19. East Wretham Heath
20. New Buckenham Common
21. Thompson Common
22. Weeting Heath
23. Ashwellthorpe Lower Wood
24. Foxley Wood
25. Honeypot Wood
26. Thursford Wood
27. Wayland Wood
28. Booton Common
29. Hethel Old Thorn
30. Hoe Rough
31. Lolly Moor
32. Sparham Pools
33. Ferry Road, Woodbastwick
34. Green Lane, Redenhall
35. Cranwich Heath
36. Backwood Lane, Brooke
37. Sherborne Road -Dersingham to Fring
38. Church Lane, Wood Dalling
39. Bedingham to Topcroft
40. St Mary's, Denton
41. Billingford Churchyard
42. All Saints, Upper Sheringham
43. Seething Church
44. Scole Churchyard
45. Little Cressingham Churchyard
46. Oxborough Churchyard

Frost in the Wensum Valley　　　　　　Photo: Nicolette Hallett

WINTER

WILDLIFE IN DECEMBER

Winter birds - How insects survive - Holly and mistletoe -
Trees, deciduous and evergreen - Tracks in the snow

'In rigorous hours, when down the iron lane
The redbreast looks in vain
For hips and haws,
Lo, shining flowers upon my window-pane
The silver pencil of the winter draws.'
From 'Winter', Robert Louis Stevenson

Seasons come and seasons go, each characterised by its own particular pattern. As winter embraces the countryside, wild creatures manage to survive the hazards of this season, as they have done from time immemorial. Just as we adjust to the changing conditions, so does wildlife - it has to if it is to survive.

Norfolk is fortunate because it has large areas which attract birds during the winter, and many different species spend some, if not all, of the winter with us. In October and November, fieldfares and redwings arrive in considerable numbers. On a cold, still, autumn evening, the calls of these birds can be heard as they make their way inland. Where hedgerows are still found, many offer a rich assortment of berries to provide a happy hunting ground for the new arrivals. Both fieldfares and redwings take insects and earthworms, together with an assortment of other invertebrates.

Many gardens attract large numbers of birds during the winter. These include starlings, the resident population being swollen as much as fourfold by visitors which have retreated from adverse weather conditions in north-east Europe. As these massive flocks of birds rise and fall, they present an almost mesmerising sight following each other in their swirling, rhythmic flights over hedge and wood, on the way to their roosting sites. If the roost is near water, the birds may take a noisy, splashing bath before they retire - depending on the temperature!

Although the coastline may seem barren and stark during winter, it is important to birds. On the sandbanks out in the Wash large flocks of the gregarious, if somewhat noisy, pink-footed geese rest, flying inland to take their fill from Fenland fields. Above the shingle ridges, snow buntings skim and gyrate, and the mudflats are alive with large flocks of knot, often thousands strong. These birds arrive from the north of Europe to winter in Norfolk's more hospitable climate. Inland, in the Breckland, the crossbills are busily feeding on the rich pickings of seeds in what seem never-ending coniferous woods.

Our own birds are still around and the robin frequents the garden, perching on a suitable object like a gate post, never more delightful than when his red breast is set against the contrasting pure white snow. Unlike many birds he keeps up his song out of the breeding season to warn would-be intruders that it his territory and he will see off any unwelcome visitors. Both song and mistle thrushes are also about. The former is well-known as it hunts its food in a business-like fashion using an 'anvil' to break open snail shells. Although the soft-bodied animal hastily retreats into the shell when the thrush approaches, the protection is not enough, and the blow of shell against stone removes the creature's refuge, providing the thrush with a tasty meal. The 'bashing stone' is used time and time again, and collections of shells are a feature of a well-used anvil.

The mistle thrush delights in sitting on the highest branch of a tall leafless tree, and from about Christmas it begins to sing. It is often called the 'mistletoe thrush', because mistletoe forms part of its diet. The winds and storms have no effect on the bird's song; indeed it has often been called the 'storm cock', as it seems that the more the wind howls, the louder he sings!

At times everything seems to be still, and life appears to have stopped in the countryside. But what a false assumption this is! When a layer of snow spreads as far as the eye can see over the Norfolk countryside, signs of life are easy to spot. The numerous footprints of fox reveal how many there are in the county, and even the 'occasional' naturalist may be able to do some tracking.

Grey squirrels leave their dreys on most days to look for food, which they secreted earlier. Otters, now returning to the county, are renowned for their playfulness, and winter adds to the delight, as

Above: Grey squirrel in typical posture

Photo: Martin Smith

Photo: Chris Knights

Below: Pink-footed geese grazing a ploughed field

Right: *Sea buckthorn*

Photo: Gillian Beckett

Below: *Cotoneaster horizontalis*

Photo: Gillian Beckett

they slide down some frozen river bank into a stream or river.

There are some groups of animals which completely disappear from the scene during the winter, and there is hardly an insect to be seen in the cold months. As the cooler days of autumn encroach on the countryside most insects decrease in number until they are no longer around. In the autumn some butterflies, including the small tortoiseshell, red admiral and peacock, feed on nectar from flowers like the Michaelmas daisy before the adults succumb to hibernation. A mild spell often wakes the sleeping butterflies, and they may be seen soaking up the sun, but they soon return to their sleepy state when the weather changes back again. Although there are species which hibernate as adults, most other butterflies and moths die off in autumn. But there should be an abundant supply next spring, because nature has planned well for these creatures. A very important stage in the life cycle of moths and butterflies is 'at rest' during the winter. When the second brood of caterpillars emerges from the eggs, the larvae feed before changing into chrysalides to pass the winter safely.

Perhaps the most conspicuous of the caterpillars is that of the privet hawk moth. Several which were placed in my care fed voraciously for a few days until they gradually became more sluggish so that any movement seemed too much trouble. Within a short time the familiar green caterpillar changed into a dark chrysalis. Inactive, but safe in an underground cell, the privet hawk moths should emerge from their chrysalides next spring - if they are not eaten by hungry moles.

On a relatively mild, sunny day a bluebottle or housefly may be stirred into a brief period of activity as the warmth lures it from its hibernatory quarters. During this time, ladybirds may also emerge briefly to sit lazily sunning themselves on some leaf or twig.

But nature is strange, and when most insects have succumbed to the lower temperatures, at least one moth becomes active. Winter moths can be seen on the wing through autumn into winter, where they are common in wooded areas and orchards. But only the male can fly; the female is wingless. With the tumbling temperatures he seems unable to cope with the cold conditions and may fly almost feebly across a road in front of the car headlights. With only rudimentary wings the female has to rely on the male finding her. It has been suggested that the female is wingless to prevent her from being blown away from the place where she pupated and where she needs to lay her eggs. And here is another mystery - why come out in the cold to lay eggs which will not hatch until next spring? Having overwintered in a cocoon in the ground, she emerges and climbs up the tree where the male locates her, probably by scent. After mating, she deposits her eggs in cracks and crevices or unopened buds. Here they overwinter, emerging in spring when there are fresh succulent leaves on the trees. Having fed for a couple of months the caterpillars make for the soil in May or June where they pupate and remain almost until the end of the year. There are great variations in the populations of winter moths from year to year. Research suggests that the number of caterpillars which survive depends on the breeding successes of some small birds, like the tits, which take these to feed their offspring, decimating the population in some years.

Of all the plants of the countryside, the holly and the mistletoe must be two of the best known during the Christmas season. The holly grows in gardens, as well as in hedges and woods, and given the right conditions, it becomes a tall, upright tree. This does not usually happen, because branches are regularly lopped off before Christmas, stunting growth. Where the holly is regularly mutilated, the new leaves have much larger spines, as do the branches closer to the ground. The higher branches are less spikey because they are relatively safe from browsing animals. Holly trees are single-sexed, and for berries to be produced, male and female trees must be close together. The pollen is carried either by the wind or by insects from one tree to another. Solitary holly trees never bear berries, because pollination cannot take place. Holly is used for decoration during Christmas, and because of its association with the Christian festival, the bright coral red berries were once called 'holy' berries. Although birds take hips from the wild rose, and haws from the hawthorn, almost as soon as they appear, they like to leave the holly berries until the first hard frost. The bright fleshy coverings attract birds, which eat them and carry the seeds many miles before disposing of them in their droppings. Many seedling hollies start to germinate where birds have taken the fruits to their favourite feeding place.

Above: *Scurvy grass*

Photo: Ron Payne

Photo: Gillian Beckett

Below: *Moss (polytrichum juniperinum)*

Holly

Mistletoe is also associated
with Christmas, and like the holly,
is firmly rooted in ancient religious
custom. Of all plants the Druids revered the mistletoe, and especially
where it grew on oak. So sacred was it that they cut it with a golden
sickle. Mistletoe is parasitic and cannot grow on its own, but needs
to feed from the tree on which it grows. Birds carry the berries to
cracks in the bark of a tree, and it is also often 'planted'. The berries
are very sticky and so slide easily over the bark of the tree; they often
end up on the underside of a branch, where they are partially
protected, and may start to germinate.

Deciduous trees lose their leaves in the autumn, whereas conifers
and evergreens keep theirs. For these broad-leaved trees the loss of
leaves means that winter will not prove too much of a 'burden'. As the
air temperature falls, so does that of the soil water, which becomes too
cold for plants to use for food-making. The tree therefore sheds its
leaves - the food factory - until more favourable conditions return in
spring. Winter is a time of rest for these trees, but evergreens and
conifers - with the exception of larch - remain clothed in green,
standing sentinel over a cold, shivering and silent countryside.

The conical shape of cone-bearing trees is better suited to cope with
heavy falls of snow. Conifers evolved before broad-leaved trees, like
the oak and ash, and large numbers of them are grown in Norfolk.
The triangular shape ensures that most of the snow slides off the
branches. Deciduous trees with their irregular shape would accumulate

snow on the leaf-clad boughs causing damage. The leaves - needles - of conifers are much smaller, the outside having a tough protective coat, decreasing the amount of water which can escape. Many conifers grow in exposed places and strong winds can pass through the small leaves without doing any harm. Although these trees do not shed their leaves in autumn, small numbers are dropped throughout the year, producing a springy carpet beneath the tree. The replacements are a soft, paler green, appearing on the ends of the twigs in early summer.

Each tree produces two types of flowers, male pollen cones and female seed cones. During the winter the seed cone, which grows at the tip of a branch, is small. Usually about May the cones stand up, and the scales open. Pollen from the male cones falls down on to and between the scales of the seed cones. The scales close, and the fertilised seeds develop inside. In the following May the scales open again so that the seeds, which have papery wings, can be set free. Caught by the breeze they are carried away from the parent plant. Empty cones often remain on a tree for several years.

For the amateur naturalist, winter is the best time to start investigations because there is plenty to look for. Birds which may be too timid to visit the garden in spring and summer are hungry now, and take almost any food which is offered. A walk in the country with a crisp, fresh layer of snow, and a nip in the air, can be one of the most rewarding excursions in any naturalist's life. Tracks in the snow may lead to sleeping animals, which can often be observed. By following other tracks, it is possible to find out what animals feed on in winter; a pile of feathers may mean that a hungry fox has had a tasty meal. Nibbled cones suggest that squirrels and mice have their winter quarters nearby. Tracking nature, and discovering her ways in winter, is often as rewarding as observations made during the rest of the year.

Cones

PLANTS IN WINTER

Under the ground and under the ice - The humble dandelion -
Scurvy grass - Mosses

During the winter months, when it appears that it is virtually at a standstill, plant life is often forgotten. Although some plants, like grass, show above the ground, growth is generally minimal, and many plants are not visible, having made adequate preparations for these difficult times. In autumn some produced seeds which remain dormant on, or in, the ground. There are others which spend this period as bulbs and corms, and yet more die off above ground level, retaining their underground stems. But in spite of their inactivity they are not dead. Although the leaves of plants like celandine may not be seen above the ground, below the surface tiny flower and leaf buds are set in very short stems, with roots below them, swollen with food which they stored earlier. These buds are ready to spring into life when they receive a 'signal'.

Soil provides some plants with protection beneath the surface, because the danger for most plants during the winter is from frost. Once the soil water temperature drops, and before it reaches freezing point, the roots are unable to take it in. This is one of nature's precautions, because the frost would turn the sap into ice, breaking the plant up from the inside. The majority of frosts only penetrate the first few centimetres of soil, and because most plants are below this, they are safe. Wind also affects plants, removing a lot of the water from the leaves by the process of transpiration. Normally this is replaced from the soil, but in wintry conditions this would not happen, and the plant would die. Most aquatic plants are also safe, because only the top few inches of the pond freezes. Those with floating leaves die down before winter. One example is frogbit. For much of the year the plant's leaves float on the surface of the water and the roots hang down. During late summer special buds which contain food are produced. In autumn these drop off and fall to the bottom of the pond, where they remain dormant until favourable conditions return. The young buds use up some of this food early in spring, making them lighter so that they can float to the surface where they begin to grow again.

During the winter months wild flowers are conspicuous by their absence although some, like the dandelion, survive. Most dandelions keep their leaves throughout the year, and in open areas the first flowers appear during warm spring days. Flowering continues, with some plants holding their golden-yellow heads high until late autumn when they die in the first severe frosts. Other dandelion plants which grow in sheltered places may flower throughout the year. Dandelions grow close to the ground, their rosette leaf formation giving them the maximum protection from activities such as trampling and mowing The leaf arrangement ensures that they do not cover each other, enabling them to receive the maximum amount of sunlight. The deeply-grooved mid-vein allows water to run down towards the centre, where it sinks into the ground providing the root with a supply of moisture. Each flower has between one hundred and three hundred florets - miniature flowers. Each head is borne on a hollow stem, containing a bitter, milky-coloured liquid.

Dandelion

Dandelions only open on fine days, and although not all florets open at the same time, none is open for more than eight hours. The first florets open on the outside, the last at the centre. In fine weather it takes about three days for all of them to open. After the fifth day the first florets fade and the head closes. Seed production only takes place if the flowers are pollinated, and in dandelions this occurs in one of two ways. Cross pollination takes place when insects carry pollen from another plant. But if this does not happen self-pollination occurs.

After pollination the florets close to allow the seeds to ripen. Each seed has a 'parachute' which serves two purposes. First, it prevents the seed from falling directly to the ground, and second, when the wind catches it the seed is carried away from the parent plant. When it lands, stiff hairs help to hold it down, and if the ground is reasonably soft, the seed may begin to grow.

Dandelions continue to grow no matter how often they are chopped down. The main tap root is similar in shape to a carrot, although smaller. Unless the whole of the root is removed it continues to grow, and even the smallest piece forms buds which push upwards to form a new plant.

The mildness or severity of a particular winter determines when plants come into flower. One plant which flowers in February, given the right conditions, is a species of scurvy grass. It has several names including early scurvy grass, because it is the first of the three East Anglian species to flower. It is also known as ivy-leaved and Danish scurvy grass. Rather rare, it is virtually confined to parts of the North Norfolk coast, where it can be found on some shingle beaches and the drier parts of saltmarshes. Fully grown plants are seldom more than six inches (15cms) in height, and it often grows along the ground. There are two types of leaves. The upper ones are generally triangular in shape, not unlike ivy leaves, hence one of its alternative names. At the base of the plant heart-shaped leaves occur. A miniature form grows in shingle and to a lesser extent on walls. Its flowers are usually white, although pink varieties are common, which is also true for the other species of East Anglian scurvy grass. It can withstand extremes of temperature and it often occurs north of the Arctic Circle.

The origin of the plant's common name is interesting. Scurvy was once prevalent and occurs when there is a deficiency of vitamin C, leading to swollen gums, haemorrhaging and general debility. It was particularly rife amongst sailors, because they did not receive adequate supplies of the vitamin. To overcome this they ate scurvy grass, the leaves of which are rich in vitamin C.

The mosses produce the most luxuriant winter vegetation, and when almost all other plants die, these seem to thrive, growing best where the atmosphere is moist all year round. Although there are few areas of true bog left in the county, there are numerous damp habitats where pockets of sphagnum moss grow in profusion. One of the most luxuriant Norfolk species is hair moss or 'urn moss', which has rather stiff and spike-like leaves. It acquired the name 'urn moss' because the capsules, borne on relatively long stalks, have lids. Other mosses grow in damp woods and on many old mole hills. Rotten tree stumps and logs are often covered with mosses, including the feather moss. In very damp woods the boughs of trees are festooned with a variety of species making rich velvet-like foliage. In untrodden areas the ground beneath trees is carpeted with moss which cushions the tread.

So life goes on for plants in winter, just as it has done during the rest of the year. Although annuals die down they have made arrangements so that their offspring appear in the following spring. As with animals, most plants have a dormant or resting stage, but they are ready to obey nature's 'command', so that once the weather is suitable, growth begins. Within a relatively short time most respond to the signal and make preparations for the days ahead.

BIRDS IN THE GARDEN

*Birds in winter - Bird-tables - Feeding birds - Suitable plants -
Nest-boxes - Respecting the privacy of birds - Bird counts*

During winter many people feed birds when they come into the garden in search of food, especially in cold weather when wild food is difficult to come by. From time to time unusual species turn up, like jays and great spotted woodpeckers. Fieldfares and redwings may also make an appearance, especially if there is some fallen fruit about. Throughout the year birds have to search for food, but winter often leads to a scarcity, and this is especially true in severe winters like 1962-63, when frost persisted for many weeks. Such winters are fortunately rare, although usually unexpected. Just as we do not anticipate them, neither does wildlife. Birds, like kingfishers, which need open water to catch live food, are unable to deal with continually frozen ponds, rivers, lakes and reservoirs. In 1962-63 the death rate among these birds was extremely high, and it was many years before they reached their earlier population levels. There is little that we can do for birds like the kingfisher, but when food becomes scarce for some other species we can help by putting out food for them.

During the extreme conditions of 1962-63, surveys suggested that only about half the bird population survived, which meant that well over 20 million birds died during the prolonged severe conditions. For one species, the diminutive wren, only one-third was thought to have survived the winter. Hard winters do not necessarily begin early, and birds may need feeding at any time. Over the years large tracts of the Norfolk countryside have been given over to development - roads, factories, housing estates - resulting in fewer areas for birds, so a bird feeding programme should be carried out throughout the year. Although this was once frowned upon, the main bird organisations now accept the necessity for this, recognising that providing food throughout the year does no harm and may even save lives. However, it does need to be carried out using recommended food.

Many people put up bird tables which attract birds throughout the year. Although Norfolk is basically rural in character, urban areas continue to increase and many people do not have extensive gardens. But with a bird table, birds can be successfully encouraged to feed. The simplest form is a wooden tray fixed to a window ledge. This will encourage many regular visitors, but where possible a slightly more elaborate bird table is better. Birds do not like strong winds, and they will not feed at an exposed table in these conditions. Strong winds also remove food from the table. Walls, hedges and shrubs afford better protection for the bird table, reducing the strength of the wind. A roofed table is worth the extra cost, because the cover keeps the food dry as well as providing shelter. A rim about 1.5 inches (375mm) around the table is necessary to stop food from being knocked off during feeding sessions, which are not always carried out with dignity!

Some bird species may remove most of the food from the table depriving the smaller ones of their food. Starlings are particularly notorious for this, and if they are not stopped then other birds are driven away. Starlings can be fed, but discouraged from using the bird table in two ways. They are generally ground feeders, and can often be seen strutting about lawns early in the morning, looking for food. If food is provided on the ground within sight of the table, they should feed here without disturbing the peace above. One inch (25mm) wire netting mesh fastened to the floor of the bird table also serves to discourage starlings as well as other larger species. Some birds prefer to feed on the ground rather than at a table, and these also need to be catered for. A tin lid is a suitable 'table' for ground feeders, but this must be brought in at night. If it is left out, remaining scraps attract undesirable rats and mice. Permanent bird tables can be fixed in a variety of positions. Apart from using window ledges they can be secured to a metre high pole. Alternatively brackets can be used to fix tables to walls of outbuildings. Some people prefer to hang their tables from branches of trees, and others use hanging basket-type feeders, especially useful where space is at a premium.

The feeding requirements of birds are as important as the siting of the bird table. Although seed-eating birds take other food in winter it is better to feed them - at least in part - on their natural diet. Fat meat is also relished by most birds, including seed-eaters. This includes the rind from ham and bacon, as well as shredded suet. Bread is

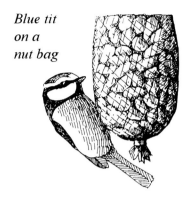

Blue tit on a nut bag

often given to birds, and brown is more nutritious than white. One of the snags of throwing food out to birds is that not all may be eaten, and checks should be made to ensure that none is left at nightfall.

Many people put out food which attracts specific bird species for the pleasure which their presence and antics bring. Peanuts are certainly the favourite food in this category and are principally put out for members of the tit family. Other birds, including jays, chaffinches and starlings, not to mention the occasional great spotted woodpecker, either devour nuts on site or take them away if they are given the chance. Once a great spotted woodpecker finds a source of food, it may come back time and time again for large, whole, shelled nuts. When other food is scarce, rotten apples are also relished, particularly by the thrushes and blackbirds, as well as the redwings and field-fares.

The seed-eaters, such as members of the finch family and sparrows, also need catering for. Specially prepared food can be bought from most pet shops, and these seeds often produce some interesting plants in the garden in the following spring. Shredded coconut should never be given because birds cannot digest this.

Any scraps, including stale bread and cake, should be kept and made into cakes and puddings - adding currants increases the nutritional value Scraps should be placed in yoghurt pots, and cool melted fat poured into the pots to bind the material. Oatmeal and boiled rice, added to melted fat, is another alternative. Once the mixture has cooled it can be cut into small pieces and pushed into the cracks in trees. Smaller birds usually manage to get some of this before greedy species like starlings take it all. Other 'cakes' can be hung up for birds to peck at. Fresh water is also important, and should be put out daily.

Another way to encourage birds into the garden regularly is to simulate the countryside, although on a smaller scale. Suitable shrubs

117

which produce a good show of berries include hawthorn (for hedging), holly, cotoneaster, viburnum (*opulus*) and berberis. Even in the harshest conditions some berries remain throughout the winter encouraging birds to come to feed. By leaving part of the garden 'wild' a natural supply of seeds is produced. In addition, plants like michaelmas daisies and sunflowers also attract birds.

Although winter brings birds into the garden, fewer people bother about these fascinating feathered creatures during the rest of the year. Nest boxes encourage birds to visit, and hopefully to settle in for longer periods. These can either be made or bought and plans for making boxes, together with a catalogue listing ready-made boxes, can be obtained from the Royal Society for the Protection of Birds, The Lodge, Sandy, Beds, SG19 2DL. Better still, why not join the Society? Some thirty different species might take up residence, but which ones come to your garden will depend on the size of 'home' provided. Different species need different-sized boxes and entrance holes, and some need open, rather than closed, boxes. Owls obviously need different boxes to blue tits. Nest boxes should be taken down at the end of the breeding season for cleaning because the nesting material often contains parasites, which might pose a health problem for next year's nestlings.

Where home-made boxes are used, the wood should be screwed together, so that the boxes can easily be taken apart. Screws should also be used for securing nest boxes to trees, posts, etc. to facilitate easy removal. A small metal plate around the edge of the hole stops other birds from making the entrance too large.

Many birds start their courtship several weeks before they nest, and a search for nesting sites may start towards the middle of March, so nest boxes should be in place well before this. Several factors need to be taken into consideration when siting the boxes. They must

Nest box

face away from the direction of the midday sun otherwise they heat up quickly. Birds provide their own nesting material since individual species have their own likes and dislikes.

If nest boxes are erected, the progress of the inhabitants wants to be followed. However, a few precautions must be taken, otherwise the birds may desert. As little time as possible should be spent near the box. Once eggs are laid it is an offence to disturb the birds or to interfere with the nest. Birds lay at varying intervals, depending on the species. Nest boxes should not be visited when the eggs are being laid, or during the incubatory period. These precautions are necessary to avoid causing the adults to leave the nest. Once the young are half-grown no more visits should be made.

The British Trust for Ornithology, whose headquarters are at Thetford, carries out an annual Garden Bird Recording Survey. Over the years the top ten garden birds have varied, but quite often the inimitable blue tit appears in every garden surveyed, and blackbirds, robin, great tit, starling, chaffinch, greenfinch, dunnock, house sparrow and the ever-increasing collared dove, are all to be found. One bird which   *Fieldfare* has given cause for concern over the last decade is the song thrush, but recent surveys seem to suggest that the bird may be regaining some of its former ground. The severity of a particular winter may bring other species into the garden. Fieldfares and redwings turn up in over ten per cent of gardens even when the weather is relatively mild, but this can rise dramatically when conditions deteriorate. It is possible that the decreasing food source in the countryside has forced it to turn to the winter garden for nourishment.

NORFOLK HERITAGE IN TRUST - TIME TO REFLECT

*Formation of the Trust - A Variety of reserves -
Financing conservation*

Unlike many industrial areas, there are still some places in Norfolk which are relatively undisturbed and unspoilt. But with an increase in the population and more time for leisure, greater use is being made of the countryside, and pressures on rural counties like Norfolk continue to rise dramatically. It is because of this that as much of the county as possible must be preserved, and not only for its wildlife, but also for its educational and recreational value. With its mainly rural character Norfolk's countryside is richer in wildlife than many other parts of the British Isles.

Serious efforts to preserve Norfolk's unique heritage began in the 1920's with the formation of the Norfolk Naturalists' Trust - now the Norfolk Wildlife Trust. Most counties now have their naturalists trusts, or trusts for conservation, but Norfolk has the distinction of having the oldest and the largest, with some 14,000 members.

The coastal areas of our county have some of the most interesting ornithological sites in the country. These include Cley Marshes on the North Norfolk coast, where more than 400 acres (162 ha) of almost undisturbed marshlands attract birds on both their outward and inwards migratory journeys. In 1926 the area was put up for sale and Dr Sidney Long, a well-known naturalist decided that he must do something to save it. Having talked persuasively to a number of wealthy friends, Dr Long was promised what he considered was enough money to acquire the Marshes at the sale on 6 March 1926. But he had competition from a number of sportsmen from the south of England who had formulated similar plans. Between them the two groups virtually monopolised the bidding in which the price went close to the ceiling which Dr Long had fixed. At one point it seemed likely that the area would go to the south of England sportsmen, but fortunately for Norfolk's birds Cley Marshes were eventually sold to him as the highest bidder.

After the purchase, a meeting was held at the George Hotel in Cley by the people who had put up the money, and they agreed that the area should be protected for breeding birds. To do this it was decided that a Trust should be formed to manage the Marshes, and on 5 November 1926 the Norfolk Naturalists' Trust was formed as a company limited by guarantee, with Dr Long as the first Secretary. There was no share capital and there were to be no profits. Cley Marshes were handed over to the new body and the first county nature reserve was set up. In spite of the lead given by Norfolk, no more trusts were formed until Yorkshire established its own group in 1946.

The Norfolk Wildlife Trust is a conservation body, interested in both plants and animals, as well as the inherent natural beauty of the county. Two international figures have stated their own beliefs on the importance of conservation. King George VI said, 'The wildlife of today is not ours to dispose of as we please. We have it in trust and must account for it to those who come after'. The late President Kennedy summed up conservation as '.... the wise use of our natural environment; it is, in the final analysis, the highest form of national thrift, the prevention of waste and despoilment, while preserving, improving and renewing the quality and usefulness of our resources'.

In 1995, almost seventy years after the formation of the Trust, the organisation manages more than 40 nature reserves in the county. These cover an area of 2,500 hectares and 10km of coastline. Some areas have been purchased outright; in other cases agreements have been reached with the owners. Other reserves, either administered by the National Trust or English Nature, are also managed by the Norfolk Wildlife Trust. These bodies work harmoniously together for the benefit of all Norfolk's wildlife and wild places. National organisations often have the money, and the local trust the expertise.

The reserves are scattered far and wide over the Norfolk countryside (see page 100) from the internationally important Roydon Common near King's Lynn, to Breydon Water near Great Yarmouth. Special facilities are available at many reserves, including observation hides and lookout points, available for everyone from the science student to the Sunday afternoon stroller.

In 1963 the Trust purchased 140 acres (57 hectares) of Roydon

Common near King's Lynn. The decision was certainly important because during research for the proposed publication 'A Flora of Norfolk' the Common was found to have one of the richest floras of any habitat in Norfolk. Plants recorded included three species of cotton grass, the only place in Norfolk where all three grow in the same reserve. One of these, hare's tail, is considered abundant, and it is the only habitat in the county where it occurs. A Nottingham research worker studying here also found much of interest, particularly with regard to the unusual level and flow of the water which allows bog and fen species to establish themselves in the same area.

The origin of the Broads, the latest National Park, has always been a matter for debate, but it is now generally accepted that they are the flooded remains of medieval peat-pits. Although many are used for boating, there are some unspoilt areas which are in the care of the Norfolk Wildlife Trust. Here bitterns, hen harriers and swallowtail butterflies find sanctuary, together with the endangered Norfolk hawker dragonfly.

Hen harrier

The largest reserve belonging to the Trust is at Hickling Broad, which has National Nature Reserve status; 715 acres (290 hectares) were purchased as long ago as 1945, and a further 500 acres (200 hectares) were leased to the Trust in the same year, with an additional 100 acres (40.4 hectares) being acquired in January 1969. Hickling Broad's greatest asset is its bird life, which includes many rare species like marsh harriers, bearded tit, black-tailed godwit and bittern. Because of the proximity to the coast it is important as a stopping off place for migrating birds. Swallowtail butterflies are some of Britain's rarest insects and they continue to breed successfully in the reserve, which is also of considerable botanical interest. Jim Vincent, the first warden, was responsible for much conservation work in the area.

On the coast, the Trust administers Holme Dunes Nature Reserve near Hunstanton. Here 550 acres (223 hectares) were either purchased, given or acquired through agreement with the owners. July and August are certainly good months for a visit to Holme. The salt-marsh is vivid with the bright mauve flowers of the sea lavender. The flowering sea buckthorn has bright orange berries in autumn which attract birds, providing a useful source of food for autumn migrants. One plant, the sea heath, reaches its northern limits in Britain here. Orchids, including both the spotted and pyramidal species, occur together with the related marsh helleborine.

Surveys back in the 1960's produced a list of 270 plants at Holme. Details of the surveys are kept in The Firs, the Trust's house within the reserve. Birds are also an interesting feature of Holme; some 200 species have been recorded in the area. Moths have also been studied and 140 species were identified on one night.

At East Winch Common nightingales can be heard, and over at Thompson Common, near Watton, occur the largest numbers of pingos in Norfolk - shallow pools which owe their origin to the end of the last Ice Age.

The greatest problem for any voluntary organisation, and the NWT is no exception, is cash, and the Trust depends very much on donations. As a registered charity, it is controlled by its Council, the members of which give up a great deal of time to manage the affairs of the organisation. Although the Trust receives a regular income from its 14,000 supporters, 40% of whom do not live in the county, extra money comes from the sale of Trust goods, donations from individuals, as well as grants from statutory bodies and industry. At some reserves reed and sedge are harvested and sold, producing a varying income, with other money coming from the sale of permits, fishing dues, rents and grants for reserves which have National Nature Reserve status. An irregular form of income is from legacies from people who have appreciated Norfolk's outstanding natural beauty.

A glimpse at a few of the reserves shows something of the diverse nature of the Norfolk Wildlife Trust. But words cannot adequately describe the value of the Trust's activities. All the work which the Trust does depends on the enthusiasm and loyalty of its members and on the hard work of its officers. We - both exiles and residents - who are true lovers of Norfolk and her countryside want to share in the work which the Trust is doing. But its ability to continue this work, and expand in the future, must depend on the interest of the inhabitants of the county. To ensure that as much as possible of Norfolk's countryside remains 'in trust' we can all share in the work. Further details are available from The Secretary, Norfolk WildlifeTrust, 72 The Close, Norwich, NR1 4DF.

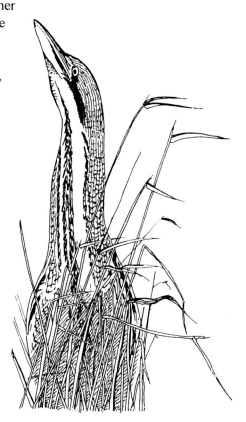

The bittern, one of Norfolk's rarest birds